Introduction to Molecular Modeling in Chemistry Education

Introduction to Molecular Modeling in Chemistry Education

Johannes Pernaa, Maija Aksela & Shenelle Pearl Ghulam

e-Oppi Ltd. & Edumendo Publishing

Publishers

e-Oppi Ltd.
Ylätie 2
31600 Jokioinen
Finland
www.e-oppi.fi / info@e-oppi.fi

Edumendo Publishing
Hiirakonkuja 7 C 15
00780 Helsinki
Finland
publishing.edumendo.fi

Cover art: Ansu Ruottinen
Cover images: alice-photo / Shutterstock.com (front), pro500 / Shutterstock.com (back)

2nd. Corrected Edition
ISBN 978-952-6649-93-1 (Paperback)
ISBN 978-952-6649-94-8 (epub)

Manufacturer: Books on Demand GmbH, Norderstedt, Germany

CONTENTS

FOREWORD

Welcome to learn molecular modeling in the context of chemistry instruction.

The goal of this book is to offer theoretical insights and hands-on activities so that chemistry teachers can implement molecular modeling in teaching.

This book includes 22 hands-on modeling exercises. They can be performed using the edumol.fi web application, which is a JSmol-JSME-based molecular modeling and visualization environment.

Users can do the exercises via any device that has a modern web browser and access to the internet. It is vital that all exercises can be done via minimal resources. Our research group has spent over a decade working with commercial software. Our experience is that schools don't have the funds to purchase equipment and software, let alone update them every three years. Free open source solutions are the only way to support the integration of molecular modeling into schools.

In this book, the level of the theory and the exercises are designed to support the work of primary, upper-secondary and high school chemistry teachers all over the world. Molecular modeling is a crucial part of chemistry instruction and chemistry educational research.

We wish you a great modeling experience with our book.
Johannes, Maija & Shenelle
Helsinki, March 2017

ACKNOWLEDGEMENTS

Our sincere thanks to Teemu Arppe for his accurate preliminary reading and hundreds of ideas how to improve the book. It was a pleasure to work with you again.

Thanks to all the graphical designers in e-Oppi Ltd. The book layout is awesome.

The book has been supported by The Finnish Association of Non-fiction Writers. Thank you for your important support.

TOPICS

The first three chapters illustrate why molecular modeling is vital in modern chemistry education. In the first chapter, we explain what molecular modeling is and why it is an important tool in chemistry instruction. How can molecular modeling support chemistry learning? The second chapter presents a technical account of molecular modeling software development. Chapter three introduces Edumol, the software used in the exercises. In the third chapter we discuss the technology and pedagogical features of Edumol.

Chapters 4–11 constitute the hands-on part of the book. The focus of the exercises is to produce concrete ideas for teaching. Many of the chapters include information regarding chemistry education: chapter 5, for example, describes informal contextual learning. Other chapters include discussions on how to model or visualize certain chemistry topics, such as isomerism, bonding, hybridization. The book ends with chapter 11, which encourages teachers to start building their own visualization exercises by providing them with the basic technical information.

1. MOLECULAR MODELING IN CHEMISTRY INSTRUCTION

Computer-aided molecular modeling which is used in chemistry research can also be an effective educational tool at different stages from basic education to universities and teacher training (e.g. Aksela & Lahtela-Kakkonen, 2001; Pernaa, 2011; Aksela & Lundell, 2008). From a chemistry teacher's point of view, computer-aided modeling is one of the most useful uses of information and communications technology (ICT) in chemistry teaching (Helppolainen & Aksela, 2015).

This chapter first examines molecular modeling, its possibilities and challenges in light of research knowledge. After the theoretical background, we will discuss the usage of molecular modeling in the planning and application of chemistry teaching. In order to understand the real world possibilities and challenges, we use a model called technological pedagogical content knowledge (TPCK).

THE POTENTIAL AND CHALLENGES OF MOLECULAR MODELING

Models and modeling are an essential part of chemistry and its teaching (e.g. Gilbert & Justi, 2016). In the research of chemistry, models are exploited at every stage of the process: forming hypotheses, observing the action of a phenomenon, explaining research results or formulating new predictions based on models. The models unite theoretical and experimental chemistry by visualizing connections between the three levels of chemical knowledge. (Justi & Gilbert, 2002). By using different models of chemistry – analogical models and computer graphics – an invisible phenomenon may be made visible, which then makes it easier to understand chemistry (Barnea, 2000). Molecular modeling is a so called metacognitive tool (Tversky, 2005) that helps us to communicate, to present information about chemistry and to process that information.

A chemistry model is used to describe a visualization of a specific phenomenon in chemistry. These visualizations help to represent thinking: they make it easier to remember and to process information, as well as to cooperate with others (Jones et al., 2005). A model in chemistry can be a concrete model like a scale model or a molecular model made of plastic, a verbal figure of speech used in an oral or a written description, a mathematical model like the general gas law, a visual model like a picture or a graph or a gestural model like the movement of a hand (Gilbert, Boulter & Elmer, 2000). Electronic molecular models can take the following forms: wire, tube, ball-and-stick, space-filling and dot surface.

Computer-aided molecular modeling is advanteageous in teaching because:

- it **helps** students understand chemistry on three different levels: macroscopic (a visible phenomenon), submicroscopic (e.g. electron density) and symbolic (e.g. a formula or a template) (Barak & Dori, 2005; Frailich, Kesner, & Hofstein, 2008)
- it **helps** to understand chemistry as a modern field of science (Aksela & Lundell, 2009)
- it **helps** to improve skills in visualization and to understand the concept of a model and three-dimensional molecular structures (Barnea, 2000, Pernaa et al., 2009)

- it **supports** the elucidation and learning of many concepts in chemistry (Aksela & Lundell, 2008; Kozma & Russell, 2005; Russell & Kozma, 2005), for example chemical bonds (Barnea, 2000; Pernaa et al., 2009), isomerism (Dori & Barak, 2001), orbitals (Flemming, Hart & Savage, 2000; Pernaa et al., 2009), functional groups (Dori & Barak, 2001), electrochemistry (Yang, Greenbowe & Andre, 2004), the structure of a substance and the proceeding of a chemical reaction (Williamson & Abraham, 1995), infrared spectroscopy (Aksela & Lundell, 2008), electron density (Aksela & Lundell, 2008), chemical equilibrium and solution chemistry (Russell & Kozma, 2005) and phenomena in biochemistry (Pernaa et al., 2009).
- it **supports** higher-order thinking skills (Webb, 2005; Aksela & Lundell, 2008; Dori & Kaberman, 2012)
- it **improves** the making of questions and the skills in inquiry-based learning and modeling, and it makes it easier to shift from three-dimensional models to structural formulas (Dori & Kaberman, 2012)
- it **inspires** towards learning concepts of chemistry (Barnea & Dori, 1996; Webb, 2005; Aksela & Lundell, 2008)
- it **supports** the handling of experimentally achieved phenomena and discussions about them (Kozma, 2003; Aksela & Lundell, 2008).

With the help of molecular modeling, it is possible to practice practice spatial skills (visual-spatial ability) that are extremely important skills in the teaching and learning of natural sciences (Uttal, Meadow, Tipton, Hand, Alden & Warren, 2013).

The challenges one might encounter when using computer-aided molecular modeling are the same as for using ICT in chemistry teaching generally: teachers usually have (i) a shortage of time allocated for molecular modeling, (ii) a shortage of technological pedagogical content knowledge (TPCK) and (iii) a deficiency of suitable software or teaching materials, and (iv) teaching large groups of students may also be quite challenging. (Aksela, & Lundell, 2008; Helppolainen & Aksela, 2015; Pernaa & Aksela, 2009)

The barriers for using molecular modeling can be divided into first-order barriers, second-order barriers and third-order barriers (see figure 1.1). For example, the necessary resources can be thought of as first-order barriers (Ertmer et al., 2012). A second-order barrier is for example a teacher's beliefs about the usefulness of the usage (Ertmer et al., 2012). Third-order barriers are design thinking skills that mean using the application at the right time in the right place in various teaching environments (Tsai & Chai, 2012). It is important to take the above-mentioned challenges into consideration in the planning and teaching of molecular models.

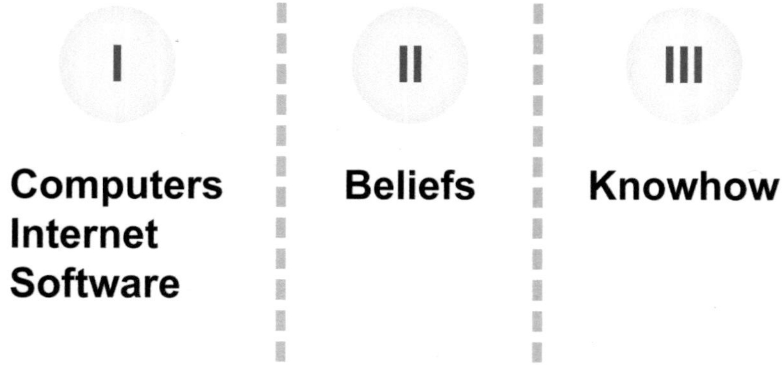

Figure 1.1 *The barriers for using ICT.*

TECHNOLOGICAL PEDAGOGICAL CONTENT KNOWLEDGE (TPCK) IN MOLECULAR MODELING

In all teaching that involves the use of ICT, including molecular modeling, a teacher needs technological pedagogical content knowledge (see figure 1.2) (Helppolainen & Aksela, 2015; Koehler, & Mishra, 2008; 2009; Chai, Koh, Tsai, & Tan, 2011; Rogers, & Twidle, 2013).

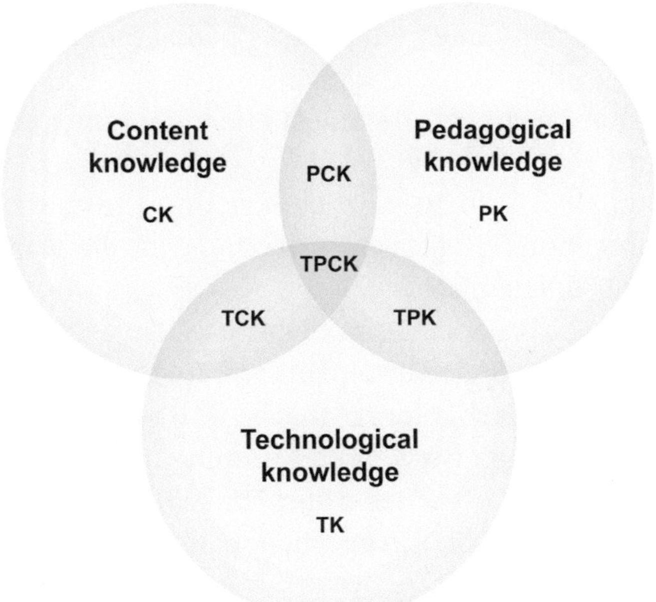

Figure 1.2 *In the instruction of molecular modeling, a teacher needs technological pedagogical content knowledge (TPCK; Koehler, & Mishra, 2008; 2009), which consists of pedagogical content knowledge (PCK), technological content knowledge (TCK) and technological pedagogical knowledge (TPK).*

TPCK describes how we can use technology in teaching, for example as a support in creating information subject-specifically (Chai, Koh & Tsai, 2013). TPCK consists of pedagogical content knowledge (PCK), technological pedagogical knowledge (TPK) and technological content knowledge (TCK). Content knowledge (CK) and technological content knowledge (TCK) have especially been observed as being significant in the development of TPCK (Chai et al., 2011). It has been observed that chemistry teachers have flaws in both TCK and TPK (Helppolainen & Aksela, 2015).

In the introduction of molecular modeling, it is important for a teacher to get acquainted with the following themes:

- information concerning the studied phenomenon (CK, the contents of chemistry)
- pedagogical information concerning the teaching of a phenomenon (PK, processes dealing with the teaching and learning of chemistry)
- technological information about molecular modeling (TK, how to use a device or a program; further information is provided in the chapters of this book). From the perspective of learning chemistry and developing the needed skills, it is crucial that a teacher uses his/her pedagogical content knowledge (PCK) in the instruction of molecular modeling.

Molecular modeling additionally contains the following areas:

- students' prior knowledge of a chemistry phenomenon studied in class
- students' attitudes and skills in using computers for learning
- information about the curriculum (aims, instruction and evaluation

- information about suitable methods for teaching and evaluation.

Technological pedagogical knowledge (TPK) refers to the various ways in which molecular modeling programs can be used to support classroom learning. In everyday practice, the teacher must adapt his or her teaching strategy to make use of the technological means that are available. For example, not every classroom has a large quantity of devices. If only one device is available (e.g. a smartphone or a computer), the teacher can use a projector to display the program for everyone and the content can be discussed together. This method works well both in student-centered and in teacher-centered approach.

If there are only a few devices available, the students can work in smaller groups. Additionally, setting up different workstations within the classroom can be a handy way to utilize molecular modeling: students can go from station to station in small groups and work on various tasks covering different phenomena. On the other hand, if every student has their own device, they can work independently at school or at home. Students can also be given molecular modeling exercises as homework.

In order to give a meaningful introduction to molecular modeling, the teacher must be acquainted with modeling programs, have technical knowledge of their performance, and be able to utilize these programs in the classroom. In this way the teacher can truly advance students' understanding of various phenomena in chemistry (TCK). The more teachers practice their technical know-how in using molecular modelling programs, the easier it gets. With the help of the numerous examples and exercises in this book, practicing cannot get easier.

STREAMLINED USING OF MOLECULAR MODELING

When teaching molecular modeling, it is important to highlight its role as a way of presenting and spreading out research results in chemistry. Additionally, it is important to discuss the restrictions of the models. This is why models need to be tested individually or in groups. Students can be divided into three different levels based on how they understand the models and how they are able to use them. First-level students think that these models represent perfect copies of the reality or that they are just "toys". Second-level students think that these models have been created for a specific purpose, but they believe these models to represent reality to some extent. Third-level students understand that these models have been created to be scientific tools. (Justi & Gilbert, 2002)

Using molecular modeling should be a part of the curriculum framework (Aksela, Lundell & Pernaa, 2008). In streamlined teaching, the aim of usage is directing the realization and evaluation of teaching. It is also important for teachers to explicitly tell students what are the learning aims of each particular lesson (see figure 1.3).

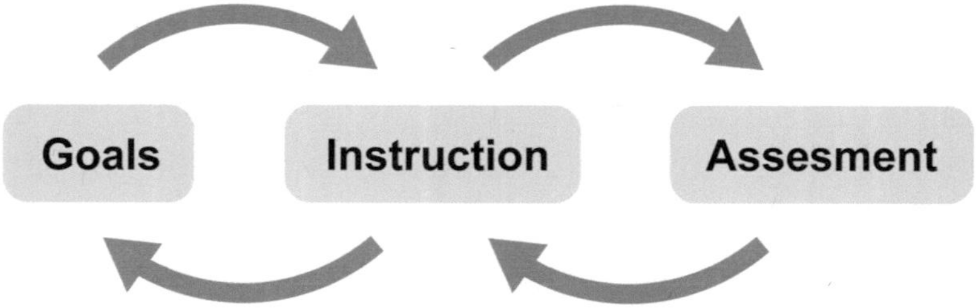

Figure 1.3 *The stages of streamlined teaching of molecular modeling.*

When using molecular modeling, it is important to make the student an active thinker and user. It is good to start using the program at an early stage. Even young children are able to practice their skills in visualization with the help of molecular modeling. In the learning of chemical phenomena, molecular modeling can be exploited starting from lower secondary school. It is also suitable for differentiation: for example, it might encourage students who do well in chemistry to develop a further interest towards learning chemistry. Furthermore, it may facilitate low-achieving students to understand chemical phenomena which are usually considered to be quite abstract. (Gilbert & Justi, 2016) Molecular modeling is also quite suitable for supporting the learning of chemical phenomena via inquiry-based activities. It can be used both before and after the inquiry-based activity, as it is used in researches in chemistry.

It is important to use versatile pedagogical teaching methods in order to support understanding of chemical phenomena. For example having a group discussion about the phenomenon before and after modeling is very important. Before starting modeling exercises, it is also advantageous to be aware of students' prior knowledge of chemical phenomenona. In order to discover what a student knows about a phenomenon, it is possible to create concept maps, have discussions or do tasks prior to modeling. Working in small groups and getting peer support may help students who experience anxiety about working with computers. Using molecular models made of plastic side by side with virtual models makes it easier for some students to understand the chemical phenomenon at hand. Combined use of new and old modelling methods makes it easier for the teacher to introduce computer-aided molecular modeling.

Molecular modeling is an important part of evaluation. Different working methods of formative evaluation (e.g. a concept map (Pernaa & Aksela, 2008), a learning diary with a self-evaluation and students' questions) support the awareness of a student during learning about achieving the given goals of learning. ICT can be used together with molecular modeling when various working methods of formative evaluation are being used. Exercises on molecular modeling are an important part of the summative evaluation of chemistry.

The role of a teacher is crucial when it comes to exploiting computer modeling in chemistry teaching (Webb, 2005; Aksela & Lundell, 2008; Pernaa, 2011). A teacher has the responsibility to direct the students' thinking and discussion, to recognize alternative concepts and to direct learning with questions.

The learning of molecular modeling and support is important from pre-teacher education to in-service teacher education. Molecular modeling is a part of teacher training, for example, at the University of Helsinki (Tuomisto, Aksela & Jääskeläinen, 2015). There is a direct correlation between the increases of subject-specific teaching being offered as part of teacher training and between the increases in the use of molecular modeling in classrooms (Rogers & Twidle, 2013; Helppolainen & Aksela, 2015). Peer mentoring and mutual discussion have been found effective in integrating molecular modeling into teaching (Rogers & Twidle, 2013). Using talented students, who are interested in using computers, as peer mentors is a good way of offering support to the teacher. In addition, peer mentoring helps students develop teamwork skills. Molecular modeling is an important tool both for teachers and students in the meaningful learning of chemistry.

LEARNING TASKS

1. Analyze the pedagogical possibilities of molecular modeling and organize them into a table.

2. Did the chapter provide new ideas and insights regarding molecular modeling in chemistry instruction?

REFERENCES

Aksela, M., & Lahtela-Kakkonen, M. (2001). Molekyylitason teknologiaa opetukseen (Molecular-level technology into chemistry education). *Kemia-Kemi, 28*(3), 200–203.

Aksela, M., & Lundell, J. (2008). Computer-based molecular modelling: Finnish school teachers' experiences and views. *Chemistry Education Research and Practice, 9*, 301–308.

Aksela, M. Lundell, J., & Pernaa, J. (2008). Molekyylimallinnuksen mentoreita kemian opetuksen ja oppimisen tueksi (Mentors for molecular modeling in chemistry education). In J. Välisaari & J. Lundell (Eds.), *Kemian opetuksen päivät 2008: Uusia oppimisympäristöjä ja ongelmalähtöistä opetusta*, (pp. 59–68). Helsinki: Universtiy press.

Barak, M., & Dori, Y. J. (2005). Enhancing undergraduate students' chemistry understanding through project-based learning in an IT environment. *Science Education, 89*, 117–139.

Barnea, N. (2000). Teaching and Learning about Chemistry and Modelling with a Computer managed Modelling System. In J. K. Gilbert & C. J. Boulter (Eds.), *Developing Models in Science Education*, (pp. 307–323). Dordrecht: Kluwer Academic Publishers.

Barnea, N., & Dori Y. J. (1996). Computerized Molecular Modeling as a Tool To Improve Chemistry Teaching. *Journal of Chemical Information and Computer Sciences, 36*(4), 629–636.

Chai, C. S., Koh, J. L. K., Tsai, C.-C., & Tan, L. L. (2011). Modeling primary school pre-service teachers' Technological Pedagogical Content Knowledge (TPACK) for meaningful learning with information and communication technology (ICT). *Computers & Education, 57*, 1184–1193.

Chai, C.-S., Koh, J. H.-L., & Tsai, C.-C. (2013). A Review of Technological Pedagogical Content Knowledge. *Educational Technology & Society, 16*(2), 31–51.

Dori, Y. J., & Barak, M. (2001). Virtual and Physical Molecular Modeling: Fostering Model Perception and Spatial Understanding. *Educational Technology & Society, 4*(1), 61–74.

Dori, Y. J., & Kaberman, Z. (2012). Assessing High School Chemistry Students' Modeling Sub-Skills in a Computerized Molecular Modeling Learning Environment. *Instructional Science: An International Journal of the Learning Sciences, 40*(1), 69–91.

Ertmer, P. A., Ottenbreit-Leftwich, A. T., Sadik, O., Sendurur, E., & Sendurur, P. (2012). Teacher beliefs and technology integration practices: A critical relationship. *Computers & Education, 59*(2), 423–435.

Fleming, S. A., Hart, G. R., & Savage, P. B. (2000). Molecular orbital animations for organic chemistry. *Journal of Chemical Education, 77*(6), 790–793.

Frailich, M., Kesner, M., & Hofstein, A. (2009). Enhancing students' understanding of the concept of "chemical bonding" by using activities provided on an interactive website. *Journal of Research in Science Teaching, 46*(3), 289–310.

Gilbert, J. K., Boulter, C. J. & Elmer, R. (2000). Positioning models in science education and in design and technology education. In J. K. Gilbert & C. J. Boulter (Eds.), *Developing models in science education* (pp. 3–18). Dordrecht: Kluwer Academic Publishers

Gilbert, J. & Justi, R. (2016). *Modelling-based Teaching in Science Education.* Dordrecht: Springer.

Helppolainen, S., & Aksela, M. (2015). Science teachers' ICT use from a viewpoint of Technological Pedagogical Content Knowledge (TPCK). *LUMAT, 3*(6), 783–799.

Jones L. L., Jordan K. D., & Stillings N. A., (2005). Molecular visualization in chemistry education: the role of multidisciplinary collaboration, *Chemistry Education Research and Practice, 6*(3), 136–149.

Justi, R., & Gilbert, J. K. (2002). Models and modelling in Chemical Education. In J. K. Gilbert, O. De Jong, R. Justi, D. F. Treagust & J. H. Van Driel (Eds.), *Chemical Education: Towards Research-based Practice* (pp. 47–68). Dordrecht: Kluwer Academic Publishers.

Koehler, M. J., & Mishra, P. (2008). Introducing TCPK. In AACTE Committee on Innovation and Technology (Eds.), *Handbook of Technological Pedagogical Content Knowledge (TPCK) for Educators* (pp. 3–29). Routledge: Oxon.

Koehler, M. J., & Mishra, P. (2009). What is technological pedagogical content knowledge? *Contemporary Issues in Technology and Teacher Education, 9*(1), 60–70.

Kozma, R. (2003). Material and social affordances of multiple representations for science understanding. *Learning and Instruction, 13*(2), 205–226.

Kozma, R. & Russell, J. (2005). Students becoming chemists: Developing representational competence. In J. K. Gilbert (Ed.), *Visualization in Science Education* (pp.121–146).

Pernaa, J., & Aksela, M. (2008). Concept maps as meaningful learning tools in a web-based chemistry material. In A. Canas, J. Novak, P. Reiska & M. Åhlberg: Concept mapping: Connecting educators – Proceedings of the III International Conference on Concept Mapping (pp. 282–289). Tallinna: IHMC.

Pernaa, J. (2011). Kehittämistutkimus: Tieto- ja viestintätekniikkaa kemian opetukseen (Design-based research: ICT for chemistry education). Doctoral thesis. The Unit of Chemistry Teacher Education, Department of Chemistry, University of Helsinki.

Pernaa, J., & Aksela, M. (2009). Chemistry teachers' and students' perceptions of practical work through different ICT learning environments. *Problems of Education in the 21st Century, 16,* 80–88.

Pernaa, J., Aksela, M. & Lundell, J. (2009). Kemian opettajien käsityksiä molekyylimallinnuksen käytostä opetuksessa (Teachers' Views of Molecular modeling in chemistry education). In M. Aksela & J. Pernaa (Eds.), *Arkipäivän kemia, kokeellisuus ja työturvallisuus kemian opetuksessa perusopetuksesta korkeakouluihin: IV Valtakunnalliset kemian opetuksen päivät* (pp. 195–204). Helsinki: University press.

Rogers, L., & Twidle, J. (2013). A pedagogical framework for developing innovative science teachers with ICT. *Research in Science and Technological Education, 31*(3), 227–251.

Russell, J., & Kozma, R. (2005). Assessing learning from the use of multimedia chemical visualiztion software. In J. K. Gilbert (Ed.), *Visualization in science education* (pp. 299–332). Dordrecht: Springer.

Tsai, C. C., & Chai, C. S. (2012). The "third"-order barrier for technology-integration instruction: Implications for teacher education. *Australasian Journal of Educational Technology, 28*(6), 1057–1060.

Tversky, B. (2005). Prolegomenon to scientific visualizations. In J. K. Gilbert (Ed.), *Visualization in Science Education* (pp. 29–42). Dordrecht: Springer.

Tuomisto, M., Aksela, M., & Jääskeläinen, M. (2015). Osaavia ja ennakkoluulottomia kemian aineenopettajia tieto- ja viestintätekniikkaa monipuolisesti hyödyntämällä (Skillful chemistry teachers through the use of ICT in chemistry teacher education). *LUMAT, 3*(6), 762−782.

Uttal, D.H., Meadow, N. G., Tipton, E., Hand, L. L., Alden, A. R., & Warren, C. (2013). The Malleability of Spatial Skills: A Meta-Analysis of Training Studies. *Psychological Bulletin, 139*(2), 352–402.

Yang, E. Greenbowe, T. & Andre, T. (2004). The effective use of an interactive software program to reduce students' misconceptions about batteries. *Journal of Chemical Education, 81*(4), 587–595.

Webb, M. E. (2005). Affordances of ICT in science learning: Implications for an integrated pedagogy. *International Journal of Science Education, 27*(6), 705–735.

Williamson, V. & Abraham, M. (1995). The effects of computer animation on the particulate mental models of college chemistry students. *Journal of Research in Science Teaching, 32*(5), 521–534.

2. FROM DESKTOP SOFTWARE TO WEB APPLICATIONS: THE HISTORICAL ANALYSIS OF MOLECULAR MODELING TOOLS USED IN FINNISH CHEMISTRY EDUCATION

This chapter discusses the historical development, current state and needs for the future development of electronic chemistry learning environments. The aim is to analyze how the requirements related to chemistry learning environments have changed over time and incorporate it into the development of mobile molecular software. The timeline of the study has been divided into three periods: the use of ICT 1) before the year 1999, 2) between the years 2000 and 2010, and 3) in 2011 and onwards. The three time periods have been created from the perspective of Finnish chemistry education, which is explained in further detail later.

For each of the time periods, the potential of one significant ICT tool in chemistry education is presented. These examples have been reflected upon in relation to the theory of blended learning and technological development in general. We argue that by understanding the history, the further needs for development can be supported more efficiently in the fast-advancing field of educational technology. Let's first take a look at the theory of blended learning in order to understand the educational needs for development.

THEORY OF BLENDED LEARNING

Blended learning is a theoretical framework that needs to be introduced in connection with computer-assisted molecular modeling. Blended learning theory guides the use and design of ICT-based learning environments in general. The theory is a bit old but still a very handy tool set for an educational ICT domain.

Blended learning is a widely studied and published topic; however, the problem is that scientists do not agree on its definition. The questions "What is blended learning?" and "What is being blended?" are the most debated topics. According to some, blending can either be viewed from the teacher's point of view (blended teaching) or the learner's point of view (blended learning). Others define the concept broadly and emphasize that blending happens on multiple different levels: blended learning is about using various types of learning activities or teaching arrangements (e.g. classroom and online learning). On the whole, according to research literature, three most commonly mentioned blending targets are:

- blending instructional delivery media
- blending instructional methods
- blending online and face-to-face instruction.

We follow Graham's (2006) definition of blended learning. According to Graham, earlier definitions are either too wide or too narrow. They contain virtually all kinds of learning and teaching activities and Graham suggests that it is more useful to define what is a blended learning system.

"Blended learning systems combine face-to-face instruction with computer-mediated instruction." (Graham, 2006)

So basically, using computer-assisted molecular modeling in chemistry teaching can be defined as a blended learning system. However, what can be achieved via ICT? Osguthorpe and Graham (2003) have summarized the key issues as follows:

1. ICT provides new media tools which engage students' interest and offer novel learning opportunities. For example, a video-based flipped classroom model, where videos are watched at home and in schools, provides more time for lab exercises and problem solving. Videos are one method of delivering media; others could be, for example, interactive molecular models with instructional questions or chemistry-related blog posts.

2. ICT enables new ways to carry out social interaction. Some students love classroom discussions; however, some students feel more comfortable demonstrating their learning in online discussion. Students' needs are diverse, and ICT provides new ways to tailor instruction suitable for various kinds of learners.

3. All this enables teaching and learning that is not restricted by time and space, if desired.

4. Lastly and most importantly, ICT has changed people's attitudes towards know-how and knowledge. Mobile devices with internet access can be brought to exams. Students have endless access to their own notes and internet databases. Therefore, do we need rote learning exam questions that only stimulate lower-order cognitive skills? Would it not be more efficient to concentrate on supporting learning skills, learning motivation and

lifelong learning? Yes it would, and this will be the future goal for learning and evaluation. Society continues to develop towards a more technology-driven direction, and more and more information is freely available. This change has already been taken into account in the Finnish education system. Problem solving through information retrieval, data processing, storing and presenting have gained greater significance in education. This also applies to chemistry education: the role of molecular models and visualization has become more vital than ever.

THE USE OF ICT IN THE 1990S

In this study, the year 1999 was chosen as the starting point of the timeline, because that was the year when the first national level chemistry education research was carried out. In 1999 all Finnish chemistry teachers were delivered a questionnaire which aimed to survey the state of Finnish chemistry education.

In the 90s molecular modeling was not used much in Finnish schools. This is explained by the fact that molecular modeling within Finnish education research and training did not begin until 2001. At that time microcomputer-based laboratories (MBL) were the most used ICT tool in chemistry education. According to the 1999 study, seven percent of teachers used MBL-tools at the time. As their motivation for using novel technology, these teachers gave the following reasons:

- Computers make data processing and presenting easy, which leaves more time for thinking.
- It enables demonstrations in large groups.
- It provides new visualization tools.
- Modern technology motivates teachers and students.

From a technological point of view, in the 90s all school computers were desktop models, internet was not common and all software had to be installed via centralized IT management. Teachers didn't have the right to install anything on school computers, which made adopting new tools very slow. However, blended learning was already present. When teachers' perceptions are compared to the central issues of blended learning, it is seen that MBL applies based on 1) new media tools and 4) data processing.

THE USE OF ICT IN 2000–2010

The new millennium brought more ICT resources for schools, and the number of computers grew rapidly. In a few years time, laptops arrived and learning began to break away from the restrains of time and place. In Finland, molecular modeling courses for teacher-education students and teachers began in 2001. Gradually Finnish literature on molecular modeling for teachers was published and some teachers began to adopt molecular modeling in their teaching.

After 2001, we have carried out a lot of research with teachers. We wanted to study teachers' attitudes towards modeling and gather knowledge about their needs. We needed this data in order to design suitable training for teachers (see e.g. Aksela & Lundell, 2008). Our research revealed the reasons why teachers wanted to use molecular modeling and what chemical topics they wanted to model. The most desirable topics were chemical reactions, chemical bonds, molecular shape, isomerism and orbitals (most of these topics are discussed in this book). At that time, the main ware being used in our courses was Spartan Student; however, Jmol was incorporated into the software selection in 2008. Jmol seemed very promising because it was free and could be used directly with a web browser without any additional installations. At that time, centralized ICT maintenance was still a major problem in schools because they often reacted too slowly to teachers' needs.

According to our research, teachers wanted to use molecular modeling for the same reasons they wanted to use MBL in the 90s. Chemistry teachers believed that modeling supports students' creativity, research skills, chemical thinking and motivation to learn. Reflecting on this, based on the theory of blended learning, it can be seen that all the key issues 1 (media), 3 (time and space) and 4 (knowledge and data tools) are present. Issue 3 (time and space) is obtained via laptops that enabled some mobility.

THE USE OF ICT FROM 2011 ONWARD

The third time cutoff is 2011. The year 2010 was a disaster for school level molecular modeling because iPads were introduced to the public. iPads were the first tablets, and problematically they did not support Java technology. At that time our research program had stopped using Spartan software, due to its relatively high price, and we had switched to using Jmol instead. However, the problem was that Jmol was a Java-based software and Finnish schools started to purchase iPads. Teachers did not understand the limitations of tablets, and many wondered what to do with these new devices. Nowadays mobile phones and tablets are the most widely used devices in schools. Therefore, it is understandable that all school-directed ICT solutions must be compatible with all mobile devices.

It took nearly two years for the Jmol developing community to adapt to this new technological situation and develop new Java-free solutions. And then it happened. The alpha version of JSmol – JavaScript-Based Molecular Viewer From Jmol was released in 2012.

The first version of JSmol was a bit slow and therefore unusable in the renewed mobile school world. It took nearly a minute to load a model. Luckily new versions were launched soon, and a more advanced version was published in spring 2014. These versions encouraged us to write this book and start working with the mobile environment.

Table 2.1 presents a comparison between all the software being used in our educational molecular modeling courses. The software comparison comprises Spartan, Avogadro, Jmol and JSmol. The comparison illustrates that each software has its strengths and weaknesses. Teachers must be aware of what they need and must choose the software according to their needs.

TABLE 2.1 COMPARISON OF VARIOUS MODELING SOFTWARE

	Spartan (1	Avogadro (2	Jmol (3	JSmol (4
Open and save work	x	x	x	x
Molecule editor	x	x	x	x
2D sketching				JSME integration
Computational methods	- Molecular mechanics - Semi-empirical methods - Hartree–fock - Density functional theory	Molecular mechanics - Ghemical - UFF - MMFF94, MMFF94s	Molecular mechanics - UFF - MMFF94	Molecular mechanics - UFF - MMFF94
Visualization tools	x	x	x	x
Image export	x	x	x	x
Animation export	x	x	x	x
3D Export		x	x	x
Mobile compatible	IOS app			Web app

1. Spartan is a modeling software published by Wavefunction Inc. It is delivered as a desktop solution for all major platforms and as a mobile app for iOS. It has different version for students, universities and business organizations. The price for e.g. high school lab pack for five students is $600 (global price). (www.wavefun.com)

2. Avogadro is an advanced molecule editor and visualizer designed for cross-platform use in computational chemistry, molecular modeling, bioinformatics, material science, and related areas. It offers high quality rendering and a powerful plugin architecture. (http://avogadro.cc)

3. Jmol is a free, open source molecule viewer for students, educators, and researchers in chemistry and biochemistry. It is a cross-platform, running on Windows, Mac OS X, and Linux/Unix systems. (http://jmol.sourceforge.net)

4. JSmol is a JavaScript-Based Molecular Viewer from Jmol. (http://wiki.jmol.org/index.php/JSmol)

This brief comparison shows that free mobile tools are more limited than commercial desktop solutions. This is obvious. Free software also crashes more often and its translations are less accurate than those of commercial software. However, the real problem is the chemistry. For example, Spartan offers ab initio calculation methods that are needed for many basic chemical problems. Free software often offers only molecular mechanics or semi-empirical calculations, which have several limitations and cannot be used in various tasks that teachers would like to perform. This book only covers molecular mechanics calculations because it is the only calculation tool that JSmol offers. It is important to remember that JSmol is, first and foremost, a visualization environment that provides a graphical interface for almost all types of file formats. So does JSmol meet the needs of teachers? Well, we honestly believe so. If we did not, then we would not have written this book and developed the Edumol application.

Finally, when this technological context is reflected upon based on blended learning, it is clear that all four possibilities achieved via the use of ICT in education are present in mobile learning environments. Firstly, mobile devices enable 24/7 access to data. Secondly, as mobile apps have interfaces to social media environments and social interaction, the time and space aspects of learning are taken to a new level (see table 2.2).

TABLE 2.2 CHANGE OF BLENDED LEARNING DURING THE THREE TIME PERIODS

	–1999	2000–2010	2011–
1. Media tools	X	X	X
2. Social interaction			X
3. Time and space		X	X
4. Knowledge	X	X	X

LEARNING TASKS

1. Carry out your personal need analysis for molecular modeling. What do you wish to achieve by using it?

2. Get familiar with some molecular modeling programs. Choose from the table above or pick another piece of software. Compare desktop and mobile molecular modeling software. Prepare a comparison of the features that they offer.

ORIGINAL RESEARCH PUBLICATION

Pernaa, J. & Aksela, M. (2013). The history, current state and future of electronic chemistry learning environments. *LUMAT, 1*(4), 435-456. **(Only in Finnish)**

The original paper was published in LUMAT. LUMAT publishes peer-reviewed research articles in math, science and technology education. It includes research papers and perspective papers.

REFERENCES

Aksela, M., & Lundell, J. (2008). Computer-based molecular modelling: Finnish school teachers experiences and views. *Chemistry Education Research and Practice, 9*, 301–308. http://dx.doi.org/10.1039/B818464J

Graham, C. R. (2006). Blended learning systems: Definition, current trends, and future directions. In C. J. Bonk & C. R. Graham (Eds.), *The handbook of blended learning: Global perspectives local designs* (pp. 3–21). San Francisco: Pfeiffer

Halverson, L. R., Graham, C. R., Spring K. J., & Drysdale, J. S. (2012). An analysis of high impact scholarship and publication trends in blended learning. *Distance Education, 33*(3), 381–413. http://dx.doi.org/10.1080/01587919.2012.723166

Hanson, R. (2012). Java-less Jmol == "JSmol". Retrieved 2015-08-22 from http://sourceforge.net/p/jmol/mailman/message/29960031.

Osguthorpe, R. T., & Graham, C. R. (2003). Blended learning systems: Definitions and directions. *Quarterly Review of Distance Education, 4*(3), 227–234.

3. EDUMOL

All of the modeling activities given later in the book are designed for the Edumol modeling environment. This chapter tells the story of the Edumol application and introduces the key features and core technologies used in it.

THE HISTORY AND CENTRAL TECHNOLOGIES OF EDUMOL

Edumol.fi is an open and free web application for molecular modeling and visualization. Its development started initially in 2009. The first version was published in the international year of chemistry in 2010. However, the usage of the first version declined very quickly, because it was built in the Java-based Jmol library. Java web solutions became outdated in 2010 due to the launch of the first iPads. iPads did not support Java web applets because of their security issues, and Jmol applets were the backbone of the old Edumol. From a software development point of view, the situation was a disaster. The new service had been at its peak of development for only a few months. Fortunately, the diffusion of iPads into Finnish schools was relatively slow, resulting in schools actually using Edumol for several years as a visualization tool via teachers' class computers.

Soon after the introduction of tablets, it became apparent that tablets would become the next big thing in schools. This accelerated the development for Jmol applet replacements, and soon new JavaScript-based solutions were published.

The new version of Edumol was published in June 2014. It took several years to develop the new version because we needed to wait for the open source community to solve all technological challenges. For non-funded projects like Edumol, open source code is the only way to produce new services. The new version of Edumol consists of:

- **3D visualization tools** that are built using the JSmol library.
- **2D sketch tools** that are build using the JSME library which is an integrated part of the JSmol.
- **A touchpad-optimized, responsive user interface** that supports Finnish and English translations. The user interface is build on JQuery Mobile tools.

JSMOL

Edumol's 3D tools are developed via JSmol library. JSmol is a JavaScript library, which enables the development of web-based Java-less interactive molecular modeling environments. JSmol is a side project of the Jmol project, which arose as a result of the proliferation of mobile devices in 2012. The project proceeded quickly and its development was carried out on a global scale immediately. People interacted via the web from various continents.

JSmol was started by professor Robert Hanson from St. Olaf College, USA. Prof. Hanson is one of the main developers of JSmol. The first version of JSmol was published via e-mail to the Jmol developers list on 2012-10-13. The email included prof Hanson's inspirational cover letter, where he described the early stages of the project, features of the first version of JSmol and a future to-do list. The new application was named JSmol. The name JSmol was derived from "Java-less Jmol". Although the work was only at its beginning, a working demo version and the pride aroused by it was truly inspirational. Take a look at Hanson's original post (Hanson, 2012) and you will see what we mean. (You can follow the link provided in the references of this chapter.)

JSME

JSME is a free and open JavaScript-based molecule editor. It has been developed to work in an HTML5 environment. JSME includes versatile molecule editor tools but, sadly, its user interface is not optimized for mobile compatible touchscreens. It is not responsive to the screen size. Despite this challenge, JSME is still a useful add-on tool for Edumol, because it enables 2D sketching from scratch and manipulation of 3D models in 2D. Additionally, it has great value for learners.

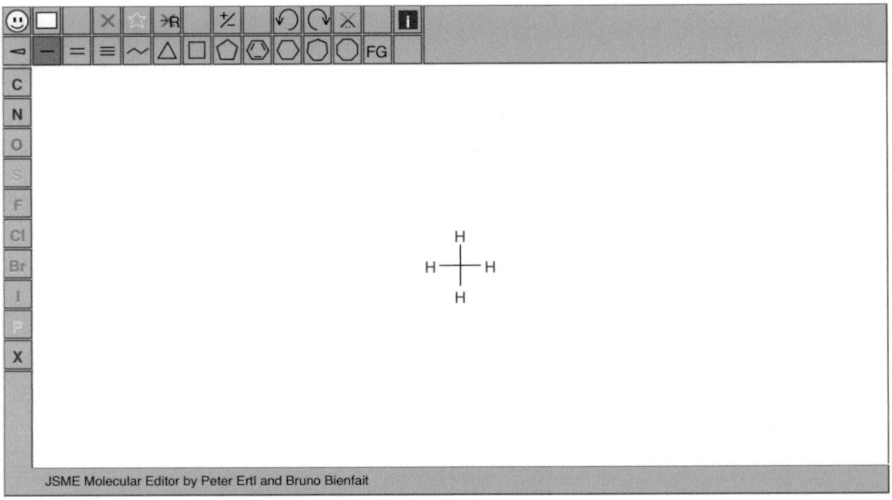

***Figure 3.1** JSME default screen in Edumol.*

EDUMOL FEATURES DESIGNED TO SUPPORT CHEMISTRY LEARNING AND TEACHING

Edumol has been designed to support the needs of chemistry learning and teaching. Its goal is to be a user-friendly molecular modeling application for all school levels and even for university students. All in all, Edumol is a highly versatile, free modeling and visualization application that combines various fields of chemistry, such as computational chemistry, cheminformatics and databases. The development of Edumol has always been oriented by educational research. This means that all the features and functions have been designed through an educational design research process. It has really been vital in the development process because Edumol only uses a fraction of all the features that JSmol offers.

Retrieving structures

Cheminformatics is the application of computer sciences to chemistry. The field of cheminformatics studies all aspects related to chemical information, for example storing, **retrieving** and presenting chemical data. Chapter 10 of this book offers Edumol exercises related to cheminformatics.

JSmol offers multiple integrations for cheminformatics. Edumol integrates three structural databases, which can be selected from the dropdown menu under the 3D canvas (see figure 3.2). Edumol uses NCI database (small molecules) by default. The database is maintained by the US National Cancer Institute, and it includes about 250 000 structures.

Macromolecules can be retrieved from the Protein Data Bank. It is noteworthy that really large structures retrieved from Protein Data Bank cannot be edited by 2D sketching. They have far too many atoms and Edumol reports this problem via an error message.

Figure 3.2 *Integrated external databases.*

Edumol also has a small internal database. It is located in the left panel. This internal database includes the example model files that are used in some exercises (see figure 3.3).

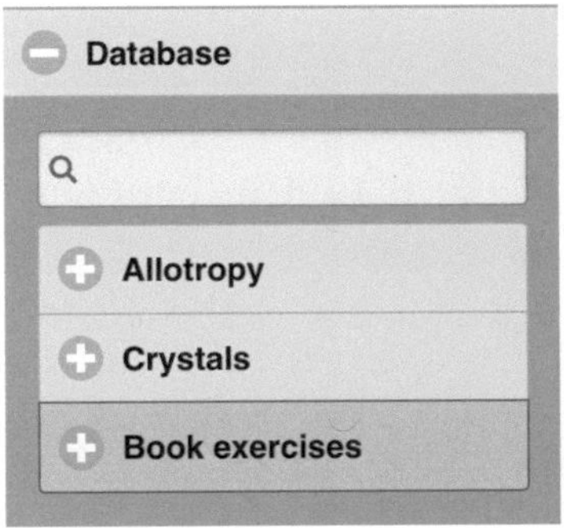

Figure 3.3 *Internal database.*

In addition to structural databases, Edumol has been integrated to NIST WebBook. Edumol can perform a NIST structural search via molecular formula, if the formula is found from the NCI database. This feature only works if NCI is up and running.

Figure 3.4 *NIST search is found on the left panel.*

2D sketching

2D sketching and 2D editing of 3D models has been considered a key feature in the development of Edumol. 2D tools are activated by clicking "3D > 2D button" on the front panel. Edumol transfers the 3D molecule to 2D editing via JSmol-JSME integration. With the 2D editor users can manipulate the model, for example, add, remove or change the atoms.

Molecular mechanics

Study the figure 3.5 from below. It visualizes different features of computational chemistry. Of the three calculation methods shown, Edumol supports the simplest one, molecular mechanics. This method causes constraints to the calculations made with Edumol.

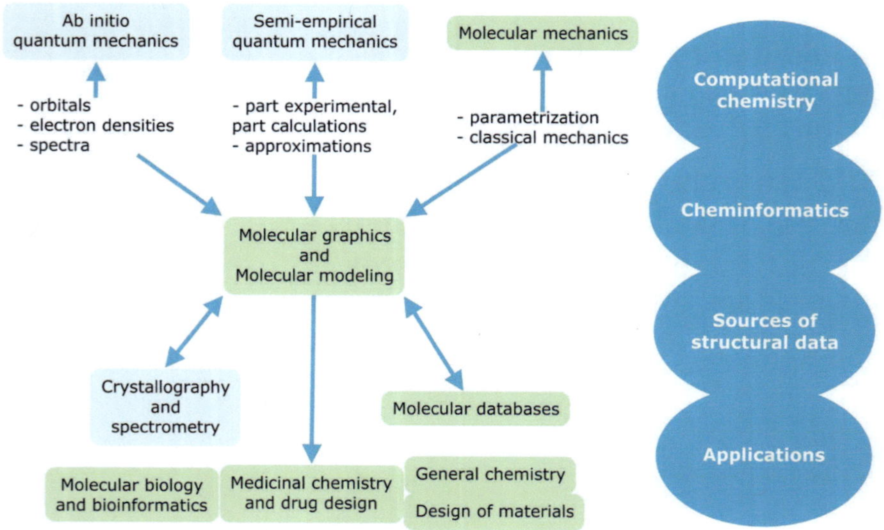

Figure 3.5 *A reflection of the features of Edumol based on the field of computational chemistry. The green boxes are covered in Edumol.*

Molecular mechanics is a computational method, which uses classical Newtonian mechanics to model molecular systems. Potential energy is calculated by considering the forces acting on the atoms. Because of this, the method is also known as a force field method.

The use of classical mechanics means that the calculations carried out are much more simpler than other computational methods (ab initio or semi-empirical methods) mentioned in the image on the previous page. In molecular mechanics, the purpose is not to calculate the electronic energy of the structure by solving the Schrödinger equation. Instead, the atoms are modeled as single particles (balls). Each particle is assigned a radius (typically the van der Waals radius), polarizability and a constant net charge (generally derived from quantum chemical calculations or experimental data). Bonded interactions are treated as "springs" with an equilibrium distance equal to the experimental or calculated bond length.

The total energy of the structure is calculated by summing up all the energy contributions that affect atoms.

$$E = E_{\text{bond}} + E_{\text{angle}} + E_{\text{torsion}} + E_{\text{plane}} + E_{\text{vdw}} + E_{\text{electrostatic}}$$

- E_{bond} is the bond stretching energy.
- E_{angle} is the energy from bending bond angles.
- E_{torsion} is the energy of the steric and electrostatic non-bonded interactions involved in the torsion of atom 4 in figure 3.6.
- E_{plane} is the energy of moving an atom out of a plane,
- E_{vdw} is the energy from the van der Waals forces between pairs of non-bonded atoms.

- $E_{electrostatic}$ describes the Coulomb interactions between partially charged non-bonded atoms.

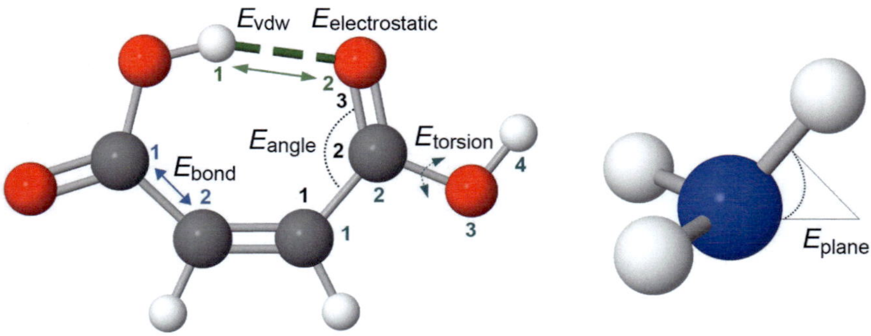

***Figure 3.6** All energy terms related to bonded and non-bonded interactions.*

The first four energy terms describe the energy related to covalent bonding. Van der Waals and electrostatic energy terms model the non-bonded interactions. They are calculated between atoms that are not interacting through a covalent interaction. Electrostatic interactions are calculated using partial charge data, so it illustrates a molecule's electric field on a rough level. Van der Waals forces are calculated using van der Waals radius values, which are derived from empirical data or quantum calculations, as mentioned earlier. Van der Waals forces simulate the interaction between electrons of different atoms. When atoms are close, they repulse each other. At an intermediate distance they are attracted, and as the distance between the atoms increases, the attraction approaches zero.

If this seems abstract, the energy terms can easily be made more concrete with Edumol by calculating the energy for some molecule and analyzing the results. In the result sheet every atom has an ordinal number which is related to the chemical symbol. This enables one to study the results of one atom at a time (see figure 3.7).

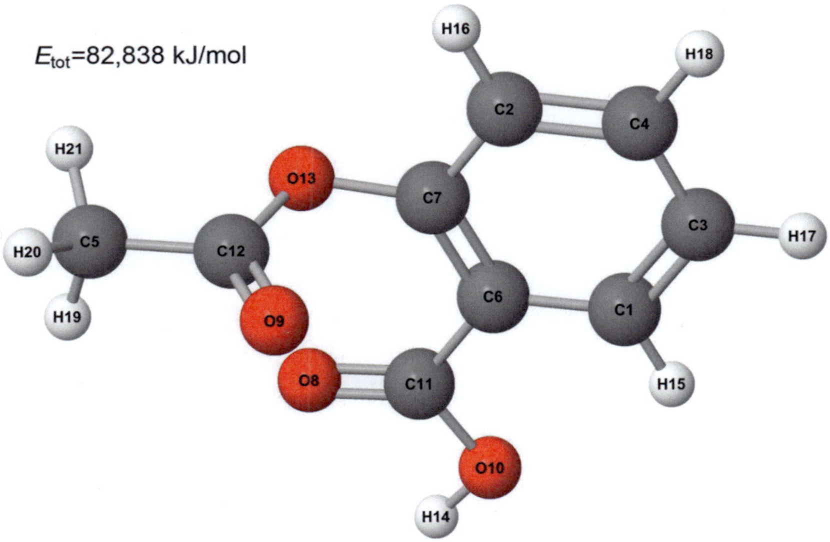

Figure 3.7 *In Edumol the atom numbers can be visualized. This enables one to examine the energy values of one atom at a time.*

All bond angles, torsion and plane energies are presented in the results sheet. Van der Waals and electrostatic forces are presented only partially. The potential energy equation can be tested by adding up all the energy term values and comparing them with the total energy (see table 3.1).

TABLE 3.1 ENERGY TERM VALUES OF ASPIRINE MOLECULE MINIMIZED USING EDUMOL

Energy term	Notation in Edumol result sheet	kJ/mol
E_{bond}	Total bond stretching energy	12.009
E_{angle}	Total angle bending energy	34.060
$E_{torsion}$	Total torsional energy	25.100
E_{plain}	Total out-of-plane bending energy	0.114
E_{vdw}	Total van der Waals energy	97.233
$E_{electrostatic}$	Total electrostatic energy energy	−85.678
E_{tot}	**Total MMFF energy**	**82.838**

Geometry optimization

As mentioned, molecular mechanics does not calculate electronic energy and the chemical world is simulated via geometric shapes: spheres and springs. This means that it cannot be used to calculate chemical reactions or orbitals; however, it can be used for optimizing molecular geometries. In the geometrical optimization process, bond lengths and angles are varied until a minimum energy structure is found (see figure 3.8).

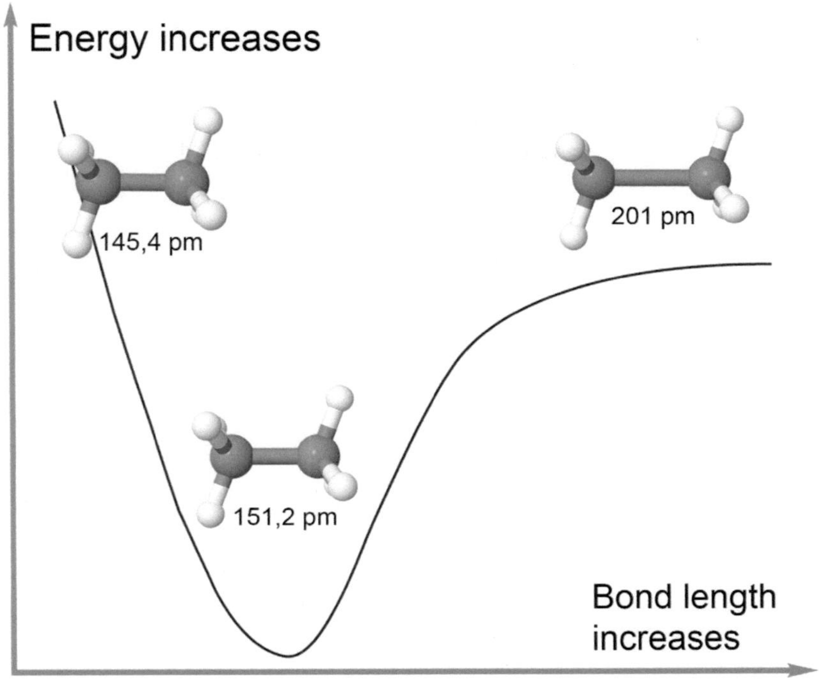

Figure 3.8 *Optimization of the bond length of ethane. The optimal bond length – not too short or too long – corresponds to the energy minimum.This figure is a rough illustration, which is why the energy values are not marked.*

Edumol carries out the minimization via the method of deepest descent, so the optimization stops at the first energy minimum. This can be the global minimum or a local minimum. Edumol doesn't take into consideration whether the minimum is local or global. However, achieving the global minimum is the goal. Hence, it is the modeler's responsibility to examine the state of the optimization process, do some adjustments and carry on the minimization if necessary. This topic is covered in chapter 9 exercise 16. The deepest descent is presented in the calculation results as follows:

```
Initial MMFF E = 131.474 kJ criterion = 0.004187 max
steps = 250

MMFF Step 10 E = 21.129844 dE = -0.133795
MMFF Step 20 E = 20.452296 dE = -0.041165
MMFF Step 30 E = 20.213366 dE = -0.020628
MMFF Step 40 E = 20.098425 dE = -0.016446
MMFF Step 50 E = 19.910973 dE = -0.008965
MMFF Step 60 E = 19.862227 dE = -0.004044
MMFF Step 70 E = 19.836314 dE = -0.002473
MMFF Step 80 E = 19.804477 dE = -0.001078
MMFF Step 90 E = 19.792480 dE = -0.000839
MMFF Step 100 E = 19.786809 dE = -0.000567
MMFF Step 102 E = 19.785671 dE = -0.000238
MMFF STEEPEST DESCENT HAS CONVERGED: E = 82.83865 kJ
after 102 steps
```

Geometrical optimization is based on the hypothesis that in nature all molecules tend to assume the lowest energy conformation. Therefore, geometrical optimization can be used to predict and validate experimental data. Further, geometrical optimization can offer support for the conclusions based on empirical measurements. Geometrical optimization via molecular mechanics can be used for studying small molecules as well as large biomolecules like proteins.

Edumol minimizes the structure mainly by using the MMFF94 force field (see Halgren, 1996). Although MMFF94 is the default force field, it switches automatically to the UFF force field, if some atoms are not recognized by MMFF94. MMFF94 has been developed for the optimization of small organic molecules. UFF is an all-round force field and it stands for Universal Force Field.

All the structures retrieved into Edumol should always be optimized because not all molecules in the NCI database are optimized. For example, optimization of the ethylene changes the H–C–H bond angle from 120° to 117.9° and the C=C bond length from 131 pm to 133.5 pm. The experimental length of the ethylene double bond is 133.05 pm. Optimization provides a more accurate value.

Hence, the function of optimization in Edumol is to produce more accurate values. In the given example the change is minor; however, the significance of optimization is further highlighted when molecules are built in 2D and then visualized in 3D. The bond angles and bond lengths of the constructed molecules can be quite far from experimental values. Accordingly, if there is a need for more accurate values, the structures should be retrieved from PubChem. Structures in PubChem are optimized via MMFF94 force field by experts. For example, the ethylene derived from PubChem has a C=C bond length of 133.4 pm.

Visualizations

Edumol modeling environment is best suited for molecular visualization because it is the core function of JSmol. Any chemical file type imported to Edumol by users can be visualized via JSmol library. Edumol can visualize various model types (see figure 3.9), molecular electrostatic potential surfaces and partial charges (chapter 6).

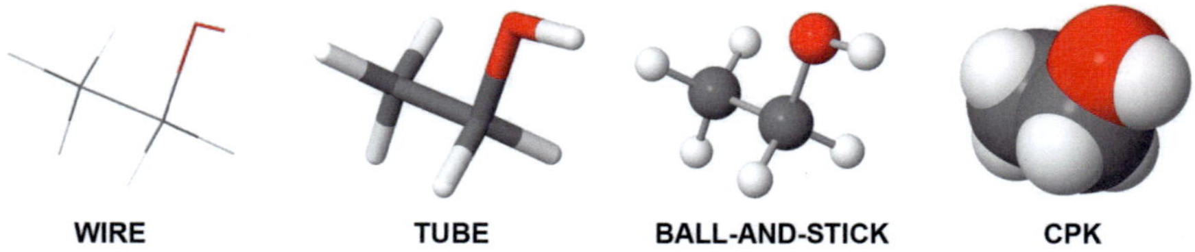

Figure 3.9 *Different model types are explained in chapter 4.*

Measurement tools

Edumol contains touch screen optimized measurement tools. Edumol can measure bond lengths, bond angles and torsion angles. Measurement tools can be activated on the right panel. (see figure 3.10).

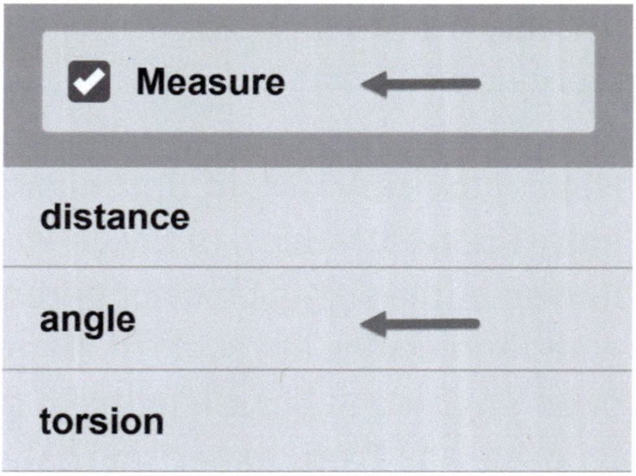

Figure 3.10 *First tap the Measure checkbox and then choose the function that is needed.*

File tools

For a web application Edumol has versatile file tools. Files can be saved and opened with Edumol, which is an unusual feature for a web application. This feature enables one to carry out more challenging visualization tasks and save the work for later. Additionally, as mentioned earlier, Edumol can read any chemical file type.

Molecules can be exported as images (PNG), structure files (MOL) or 3D models (X3D). Additionally, images can be exported as a working size or in 2000 × 2000 pixels. HQ images are saved as transparent PNG image files, which makes it easier to use them in PowerPoint applications or other presentation documents. MOL files can be opened in other molecule editing software such as ChemSketch or MarvinSketch. Furthermore, virtual 3D models can be printed as concrete 3D models using 3D printers. 3D printing is something that we will be concentrating on in the near future.

Figure 3.11 *Edumol File menu.*

Social interaction

Edumol is designed to be a social service, which provides the user with the opportunity to network and to get support. Edumol has a Facebook page, Flickr profile and Sketchfab account. The Facebook page is for news and discussions, Flickr is for images, and Sketchfab is for 3D models. The Flickr and Sketchfab profiles also support discussion and networking.

Links to social media:

- https://www.facebook.com/edumol.fi
- https://www.flickr.com/photos/edumendo
- https://sketchfab.com/edumol.

LEARNING TASKS

1. What does it mean when molecular modeling and, especially, geometrical optimization is referred to as "the link between experimental and theoretical chemistry"?

2. What are the most important features of Edumol application that you are going to use in your teaching?

REFERENCES

Bienfait, B., & Ertl, P. (2013). JSME: a free molecule editor in JavaScript. *Journal of Cheminformatics, 5*(1), 24. http://doi.org/10.1186/1758-2946-5-24

Bladon, P., Gorton, J. E., & Hammond, R. B. (2012). *Molecular Modeling: Computational Chemistry Demystified.* Cambridge: RSC Publishing.

Halgren, T. A. (1996). Merck molecular force field. I. Basis, form, scope, parameterization, and performance of MMFF94. *Journal of Computational Chemistry, 17*(5–6), 490–519.

Hanson, R. (2012). Java-less Jmol == "JSmol". Retrieved 2016-07-20 from http://sourceforge.net/p/jmol/mailman/message/29960031.

Jmol. (2015). Jmol: an open-source Java viewer for chemical structures in 3D. Retrieved 2016-07-20 from http://www.jmol.org.

Mattila, K. (2008). Voimakenttämenetelmät ja molekyylidynamiikka (Force fields and molecular dynamics). In J. Blomqvist & T. Kurten (Eds.), Ongelmasta ratkaisuun – CSC:n kemian mallinnusopas (pp. 16–26). Espoo: CSC – Tieteellinen laskenta Oy.

National Cancer Institute. (2016). About. Retrieved 2016-07-20 from https://cactus.nci.nih.gov/ncicadd/about.html.

NCBI. (2015). The PubChem Project. Retrieved 2016-07-20 from https://pubchem.ncbi.nlm.nih.gov.

Pernaa, J., & Aksela, M. (2011). Kehittämistutkimus: Kemian kouluopetukseen soveltuvan molekyylimallinnusympäristön kehittäminen (Design research: Designing a molecular modeling environment for school chemistry). In M. K. Aksela, J. Pernaa & M. Happonen (Eds.), *Kansainvälinen kemian vuosi: Kemia osaksi hyvää elämää (International Year of Chemistry: Chemistry as a part of good life)* (pp. 110–121). Helsinki: Centre for Chemistry Education. (Only in Finnish)

RCSB PDB. (2015). RCSB Protein Data Bank - RCSB PDB. Retrieved 2016-07-20 from http://www.rcsb.org/pdb/home/home.do.

The jQuery Foundation. (2015). jQuery Mobile. Retrieved 2016-07-20 from https://jquerymobile.com.

Wild, D. (2013). *Introducing Cheminformatics.* Self-published. http://learncheminformatics.com

4. COMPUTER-BASED MOLECULAR MODELS

Chemistry is a strongly visual science, which poses challenges to learning. In chemistry textbooks molecules are often represented as 2D diagrams, but to understand chemistry students need 3D models. Before the existence of molecular modeling software, students had to be able to mentally visualize molecules in 3D. This skill is called spatial ability. Spatial ability consists of three levels of knowledge

1. **Spatial visualization**, the ability to visualize 2D representation as a 3D model.

2. **Spatial orientation**, the ability to imagine how the molecule would look like from another perspective.

3. **Spatial relations**, the ability to visualize spatial operations like rotations and reflections.

According to chemistry educational research, a strong correlation has been found between spatial ability and chemistry learning. Students with poor spatial visualization skills have difficulties understanding visual topics in chemistry. For example, the Fischer projection is a 2D representation of carbohydrates. Its purpose is to highlight atom connectivity; however, it ignores the spatial relationships of atoms, which are visible in the ball-and-stick model (see figure 4.1). Some students find it difficult to understand the Fischer projection because the three-dimensional nature of the molecule is embedded in a 2D diagram. Therefore, understanding it requires spatial visualization skills.

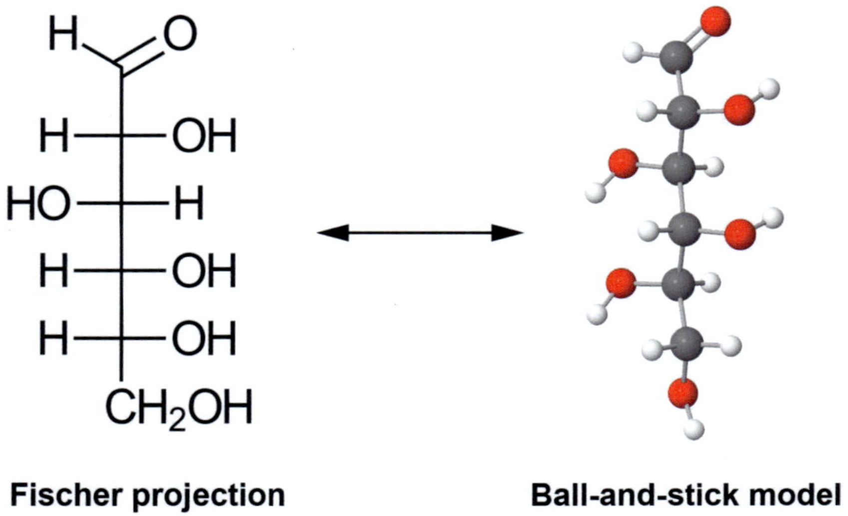

Figure 4.1 *The Fischer projection in 2D and 3D.*

SUPPORTING VISUALIZATION SKILLS: MODEL TYPES AND ROTATION TOOLS

The use of molecular modeling software supports the development of spatial visualization skills. Molecular modeling software enables the rotation of molecule models, which supports the development of spatial orientation and relation skills. Moreover, spatial visualization skills can be supported via different model type representations.

Model types

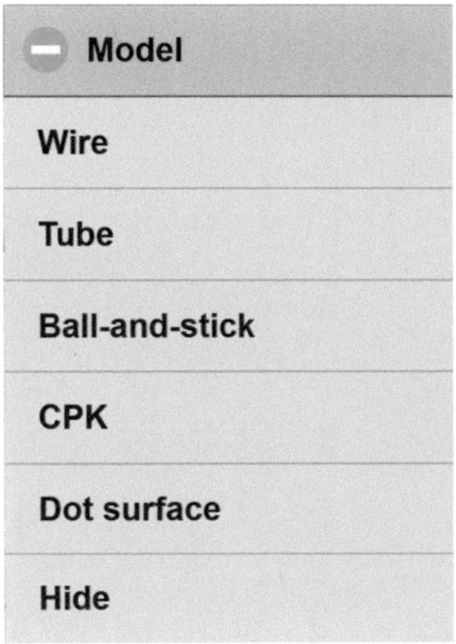

In Edumol application users can choose to use different model types according to their visualization needs. Edumol offers five model types for molecules and a hide option.

Wire and tube

Wire and tube models visualize bonds but not the size of the atoms. Bonds are represented as colored cylinders, using default atom colors. For example, a bond between carbon atoms is gray, and the bond between carbon and hydrogen atoms is drawn in two colors because hydrogen atom is colored white. These models also visualize multiple bonds (see figure 4.2).

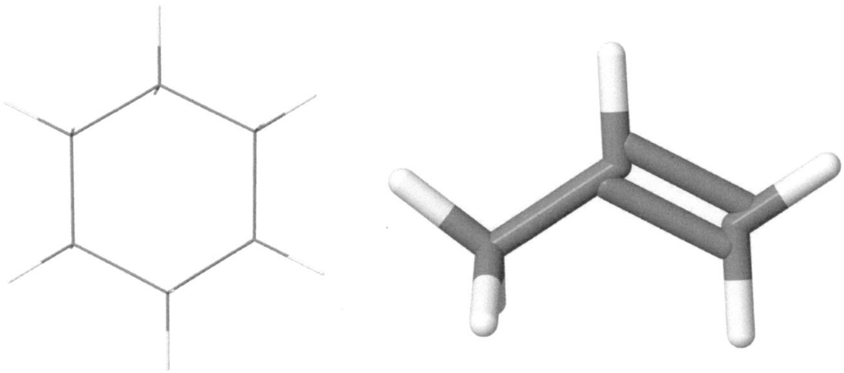

Figure 4.2 *Cyclohexane represented as a wire model and propene as a tube model. Observe the bond colors.*

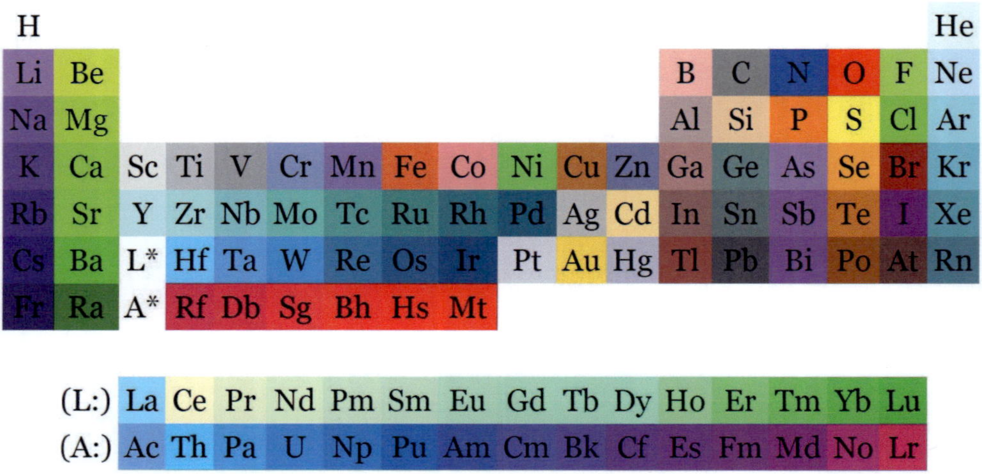

***Figure 4.3** Edumol uses default JSmol colors.*

Ball-and-stick

The ball-and-stick model is the most commonly used model. It is widely used because it displays the 3D positions of the atoms and the bonds between them simultaneously. It is the virtual equivalent of the concrete molecular representations made of plastic (see figure 4.4).

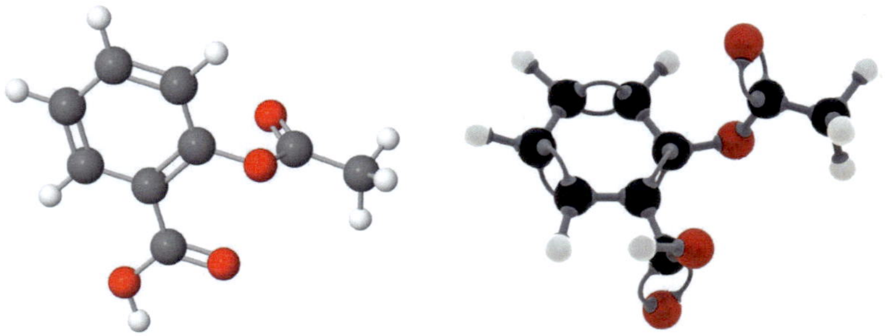

***Figure 4.4** Aspirine molecule visualized as a ball-and-stick model and a plastic molecular model (Radu Razvan / Shutterstock.com).*

In the ball-and-stick model, the radius of the atom sphere is much smaller than the length of the bond cylinder to ensure a clear view of the bond. As a result, the model gives an incorrect view of the space between the atoms.

The first ball-and-stick molecular model was made by Wilhelm von Hofmann, a famous chemist. He used such models to support learning in his chemistry lectures ca. 1860.

CPK model

The CPK model is a molecular visualization that uses van der Waals radii in displaying the atom balls. The CPK model is useful for visualizing the effective shape and relative dimensions of the molecule. However, the CPK model also has limitations. It doesn't display bonds or structures under the first atom layer. For example, take a look at this cyclizine model, in which one of the nitrogen atoms is barely visible (see figure 4.5).

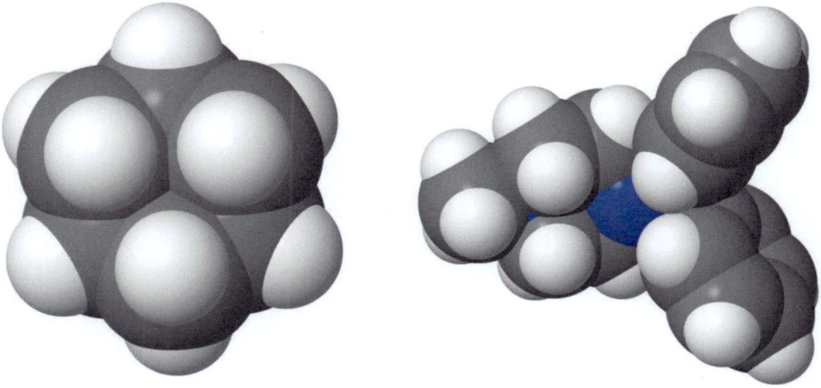

Figure 4.5 *CPK models of cyclohexane (left) and cyclizine (right).*

CPK models are also known as space-filling models or calotte models. So why does Edumol use the term CPK? CPK was named after the famous the chemists Robert Corey, Linus Pauling and Walter Koltun, who were the early users of this model type. Edumol uses the term CPK in order to honor the achievements of these great chemists.

Dot surface

The dot surface model is a combination of the CPK model and the ball-and-stick model. It displays the shape of the molecule as a dotted surface; however, the atoms and bonds are still visible.

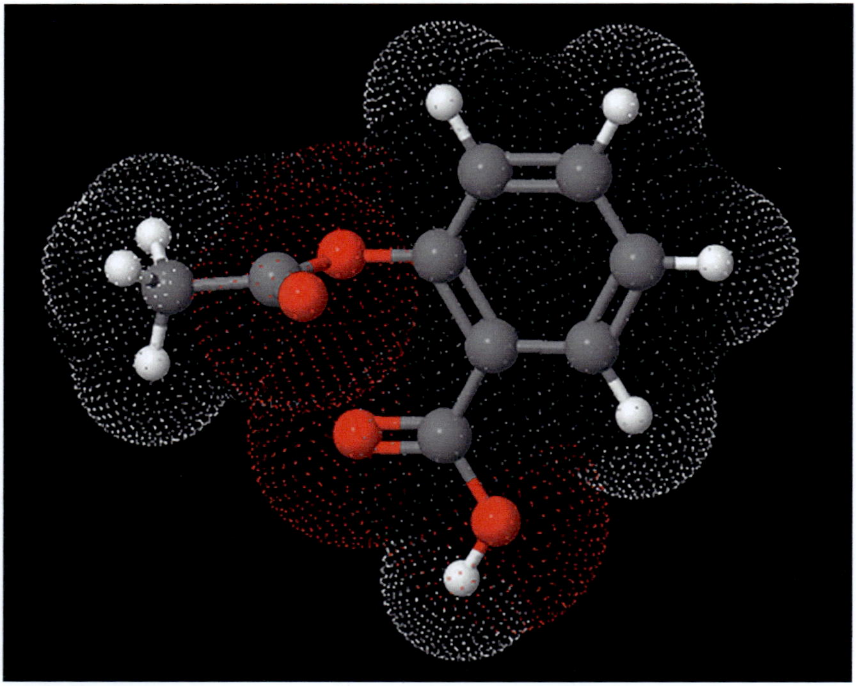

Figure 4.6 *Dot surface model looks best in a black background.*

TABLE 4.1 SUMMARY OF MODEL TYPES AND THEIR LIMITATIONS

Model	Displays	Limitations
Wire	bonds	Atoms are not displayed.
Tube	bonds	Atoms are not displayed.
Ball-and-stick	3D shape, atoms, bonds	3D space is not accurate.
CPK	3D space, 3D shape	Bonds are not visualized.
Dot surface	3D shape, 3D space, atoms, bonds	Multiple models merged in one meta level model. Requires high-level visualization skills to use.

EXERCISE 1. MODEL TYPES

Let's get hands on with Edumol. The application is very easy to use. Start by taking any device a phone, a tablet, a laptop or a desktop computer and navigate to www.edumol.fi. Choose some basic organic molecules such as ethane, ethanol, acetic acid and then start testing the web app.

The figure 4.7 displays the initial view after downloading the page. Edumol downloads the methane model onto the 3D canvas automatically. You can get another molecule model from the database by writing the name of the molecule in the text entry **(1)** and then hitting enter or pressing the Search / !Run button

(2). If you want to sketch something from scratch, Edumol also has a 2D sketch tool integrated. You can switch from the 2D side to the 3D side by pressing the button labeled 3D > 2D **(3)**.

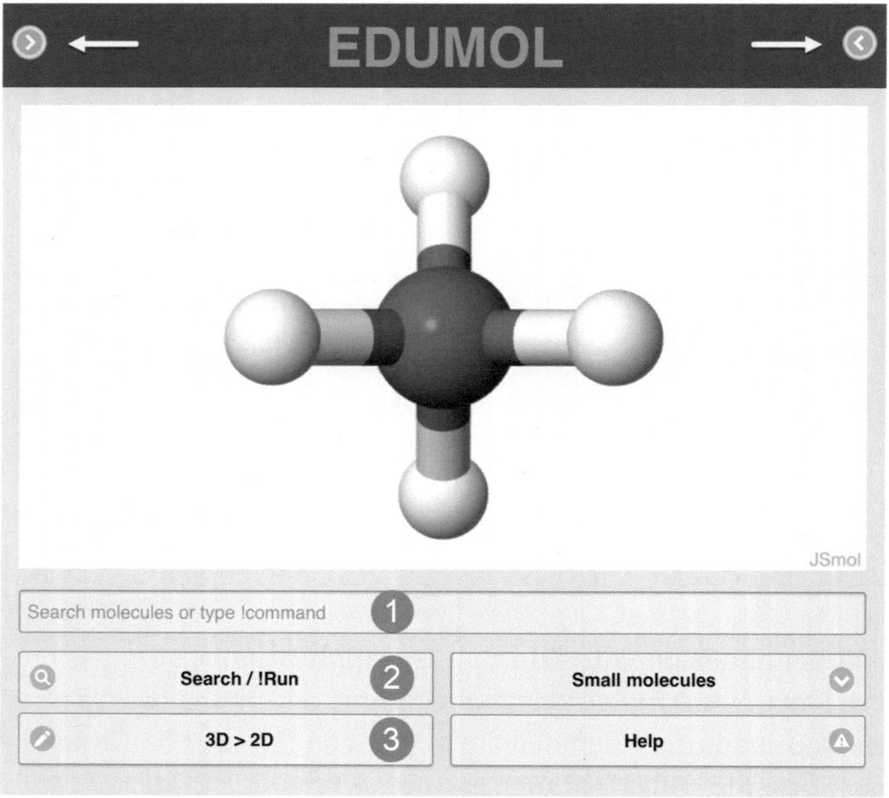

Figure 4.7 *Edumol user interface.*

After choosing (or designing your own) molecule models in Edumol, open the Model Options tool from the right side arrow button. Try out different model types and reflect on their features in the summary table 4.1.

EXERCISE 2. ROTATION TOOLS

Rotation tools are found on the right side panel under the Move menu. Molecules can be rotated to the front (Z axis), to the right (X axis) or up (Y axis).

Test your spatial orientation skills. Retrieve an aspirin molecule, for instance, and check the Axes checkbox. First imagine what the model will look like from another angle, then rotate the molecule into that viewpoint in order to test your spatial orientation.

REFERENCES

Barnea, N. (2000). Teaching and Learning about Chemistry and modelling with a Computer managed modelling System. In J. K. Gilbert & C. J. Boulter (Eds.), *Developing Models in Science Education* (pp. 307–323). Dordrecht: Kluwer Academic Publishers.

Stieff, M., Bateman, R. C., Jr., & Uttal, D. H. (2005). Teaching and learning with three-dimensional representations. In J. K. Gilbert (Ed.), *Visualization in Science Education* (pp. 93–118). New York, NY: Springer.

5. MOLECULAR VISUALIZATIONS AND INFORMAL CHEMISTRY LEARNING

Chemistry visualizations are everywhere in the media. One approach to integrate real life into chemistry learning is to use different media resources as learning material. In the educational field, this chemistry educational approach is called **informal learning**. In this chapter, we take a closer look at informal chemistry learning and suggest some modeling activities.

We use organic compound groups as our exercise topic because the study of compound groups is an extensive part of upper-secondary chemistry education and a part of chemistry curricula worldwide. Organic compound groups are also a good introductory topic for the book's hands-on part because there are dozens of familiar, small organic molecules that can be studied with students.

INFORMAL LEARNING IN CHEMISTRY

Informal chemistry education can be defined as learning that takes place outside of exam-oriented, structured school systems. Informal learning takes place, for example, when watching television, visiting museums or science parks, and when reading books, magazines or newspapers. Even if informal learning occurs outside of the school environment, it can still be a part of the course curriculum. However, its implementation needs to be carefully planned.

According to Stocklmayer & Gilbert (2002), successful informal learning includes three key aspects:

- situation
- context
- narrative.

Informal learning must be connected to a real life **situation**, like in medicinal chemistry. Medicinal chemistry transforms into situational **context** when the learner finds something that they can relate to in their personal life experiences, for example, medicating a disease by using a common medicine. The learner constructs a **narrative** from the context. The narrative is a story or a metaphor which the learner uses to explain and illustrate ideas and experiences. The learning experience is then evolved into a meaningful communication tool.

Hence, the idea behind informal learning is to facilitate positive attitudes towards chemistry learning. This is important because it is common knowledge that chemistry is viewed as an abstract, difficult and complex subject (see chapter 1). Fortunately, there are numerous interesting themes for informal learning in chemistry.

Some examples:

- **Making chemistry equations more concrete** by getting familiar with the real industrial processes that are behind them. This can be carried out via field trips, books or television shows. Chemistry is not all about symbols: it includes people, businesses and social interaction.

- **Molecular gastronomy** (food chemistry) can serve as a familiar situation for illustrating the link between macro and sub-micro levels and for practicing argumentation. This is an interesting theme, which offers several books that are suitable for students to read.
- And then there is the topic that we have chosen, organic compound groups and their characteristic functional groups. In the next sub-chapter, we present two exercises that can facilitate teachers' use of informal learning as a chemistry educational approach.

EXERCISE 3. ORGANIC COMPOUND GROUPS IN MAGAZINES AND NEWSPAPERS

The goal of this exercise is to show students how chemistry is presented in magazines and newspapers. In practice, students can analyze how organic compounds or compound groups are presented or mentioned in articles. They can then produce a brief summary, and attach representations of the molecules mentioned in the article.

Preparations: Teachers can prepare a reading list for the students in advance, by choosing a couple of familiar local newspapers or science magazines.

Example student task

The task for students can be formulated as follows:

Read the magazine or article X and write a short summary of the article. The summary should include a discussion on how organic chemistry and the organic compound groups were presented in the article.

Additionally, discuss which functional groups related to compounds or compound groups were mentioned in the paper. Then, using Edumol, visualize the molecules or functional groups that were mentioned in the article and attach the images to the summary.

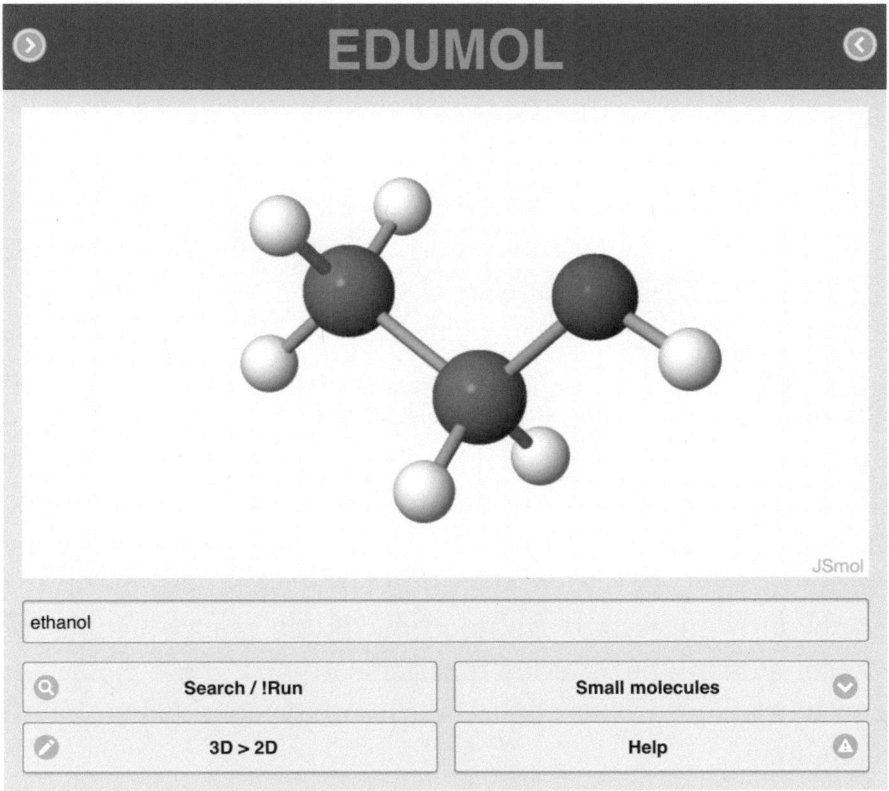

Figure 5.1 *To search molecules, type the name of the molecule into the text field and press Search button.*

EXERCISE 4. INFORMATION RETRIEVAL

In this exercise we introduce the Young Scientists Journal, which is an excellent student-based science journal that can be used to support informal chemistry learning. The informal theme of the task is antibiotics and bacterial resistance.

Journal description: *"The Young Scientists Journal is run by a group of 12–20 year olds around the world, in contact over the internet. It was started at The King's School Canterbury and a team meets there to work on the journal on a weekly basis. We also have contributors from other schools, and have a growing international community."* (Young Scientist Journal, 2016)

Important links for the teacher:

- journal homepage: https://ysjournal.com
- teacher instructions: https://ysjournal.com/about-us/information-for-teachers.

Teacher tips

The task is to search the web for the given article, read through it, and explain which functional group is responsible for the antibiotic efficacy and how antibiotics work. Additionally, the paper introduces four common antibiotic groups which are used in identifying different functional groups. These molecules can be retrieved from the database or sketched using the 2D sketchpad.

The four antibiotic groups are:

- Penicillins | Search words: “Penicillin” or “Penicillin G” (see figure 5.2)
- Cephalosporins | Search word: “Cephalosporin” (see figure 5.3)
- Monobactams | Search words: “Monobactam” or “Tigemonam” (see figure 5.4)
- Carbapenems | Search word “Carbapenem” doesn’t give any results, but e.g. “Imipenem” works. Imipenem is a β-lactam antibiotic, and its name has to be found via other sources. This last one is a bit more challenging (see figure 5.5).

For example the following functional groups are present in the above-mentioned molecules:

- amide
- amine
- benzene ring
- β-Lactam (cyclid amide)
- carbonyl
- carboxyl
- ester
- hydroxyl
- and groups containing sulfur.

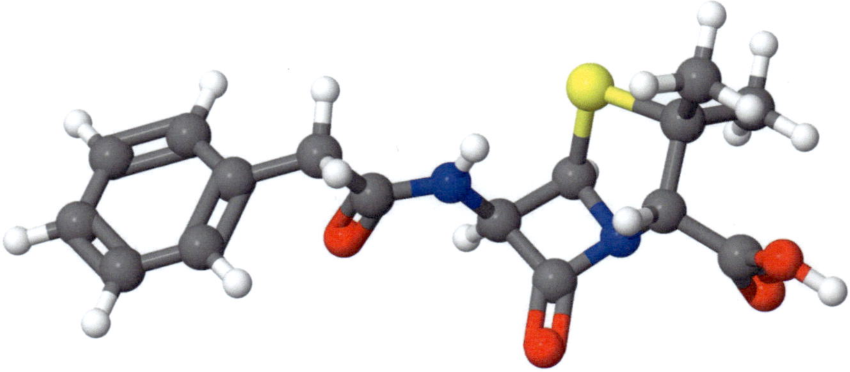

Figure 5.2 *Penicillins – Penicillin G.*

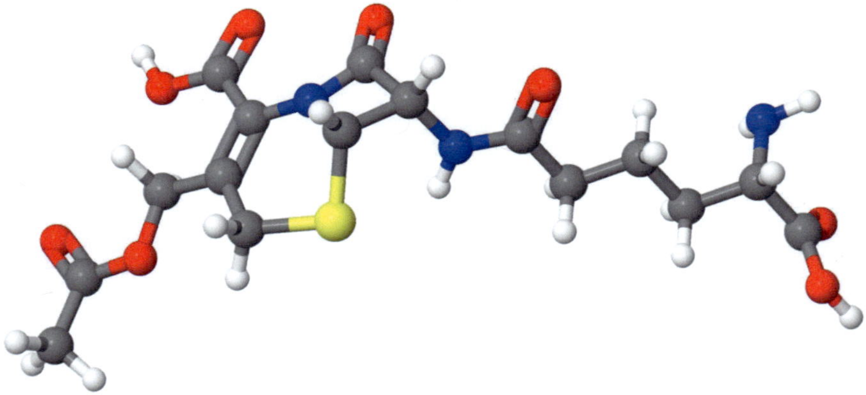

Figure 5.3 *Cephalosporins – Cephalosporin.*

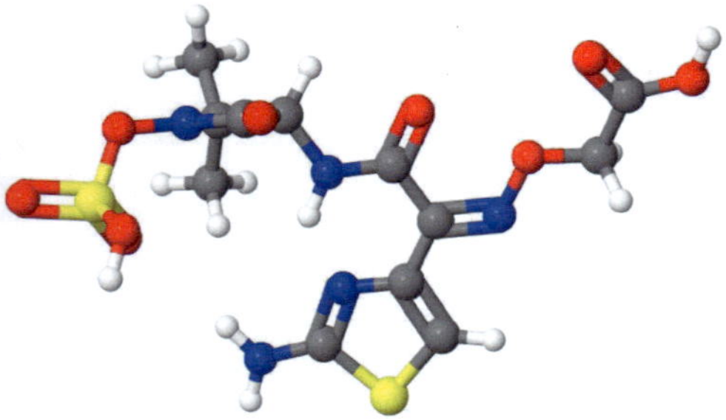

Figure 5.4 *Monobactams – Tigemonam.*

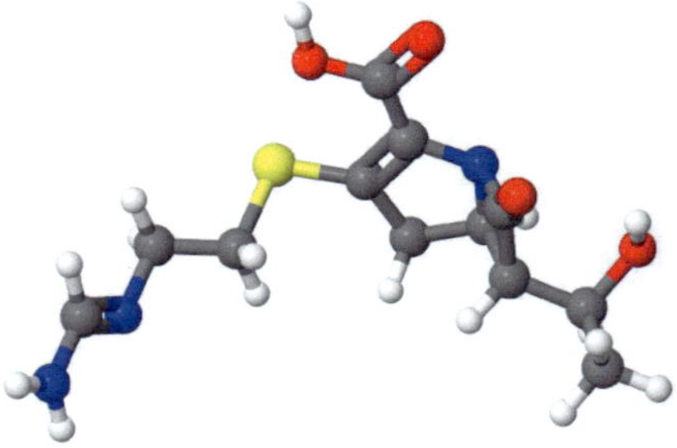

Figure 5.5 *Carbapenems – Imipenem.*

Example student task

Read the following article: Paul-Enguerrand Fady. (2014) A Comparative Study of Experimental Methods to Combat Antiobitic Resistance. *Young Scientist Journal, 15*, 8–15.

- Find out which functional group is responsible for the antibiotic efficacy and how antibiotics work.
- Search for the four common antibiotic groups that are introduced in the paper. Retrieve the characteristic molecules from the database that Edumol uses or draw them using the 2D sketchpad.
- Which functional groups are present in the molecules? Visualize various functional groups, name them and export images of them.

REFERENCE

Stockmayer & Gilbert (2002). Informal chemistry education. I In J. K. Gilbert, O. De Jong, R. Justi, D. F. Treagust & J. H. Van Driel (Eds.), *Chemical Education: Towards Research-based Practice* (pp. 143–164). Dordrecht: Springer.

6. PARTIAL CHARGES AND ELECTROSTATIC POTENTIAL SURFACE

Partial charges are atomic charges resulting from the asymmetric distribution of electrons in chemical bonds. Partial charge models illustrate the molecule's electric force field. It shows which atoms are more **electronegative** and which parts of the molecule are electron rich or electron poor (see figure 6.1). This can be used as a tool for explaining e.g. polarity, solubility and reactivity.

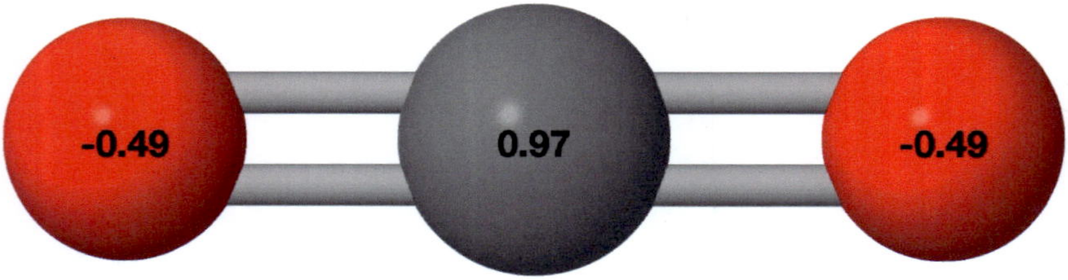

Figure 6.1 *The oxygen atom is more electronegative than the carbon atom; therefore it has a negative partial charge.*

Edumol calculates partial charges using the MMFF94 charge model. According to JSmol software documentation, multiple bonding (e.g. double or triple bonding) is required for charge calculations. This requirement seems to apply only for hydrocarbons because Edumol cannot generate partial charges for alkanes. However, partial charges can be calculated for polar molecules such as water, even when water molecule does not include multiple bonds (see figure 6.2).

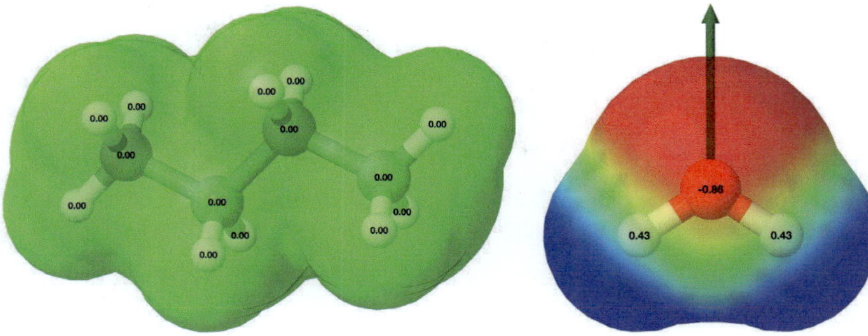

Figure 6.2 *Compare the visualizations. Edumol cannot generate partial charges for butane.*

In Edumol, partial charges can be visualized as a molecular electrostatic potential surface (MEP). MEP surfaces illustrate the charge distribution as a highly visual 3D model. It is used to explain charge-related molecular properties, and it also illustrates the shape and size of the molecule.

The figure 6.3 shows the partial charges and the transparent MEP surface of ethanol. The oxygen atom has the most negative partial charge value and is visualized as red. The oxygen atom also has the highest electronegativity value, and the electron density is highest around it. Red and blue indicate the lowest and the highest electrostatic potential energies, respectively.

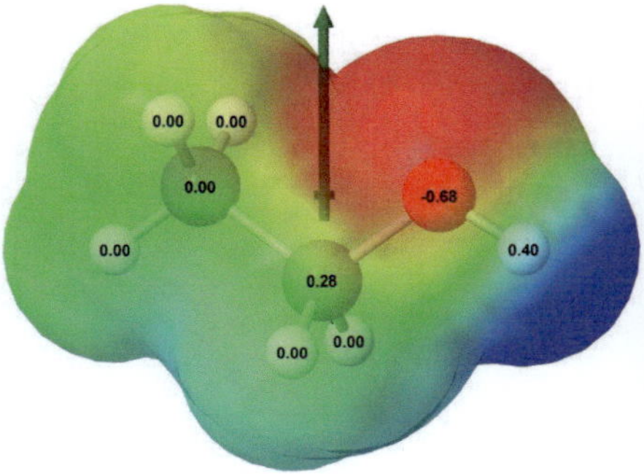

Figure 6.3 *The MEP surface of ethanol. The green dipole vector illustrates the magnitude of the electric force field and the direction from positive to negative.*

Let's analyze the figure 6.4. The hydroxyl group is the reactive site of the alcohol. The hydroxyl group has both a positive and a negative charge region. The negative area (red) has the lowest potential energy and it would have a strong interaction with positively charged particles nearby. The oxygen in the hydroxyl group is acting as a nucleophile and it donates a pair of electrons to the electrophile in order to form a chemical bond. Similarly the positive part (blue) of the OH group would have an interaction with negatively charged particles.

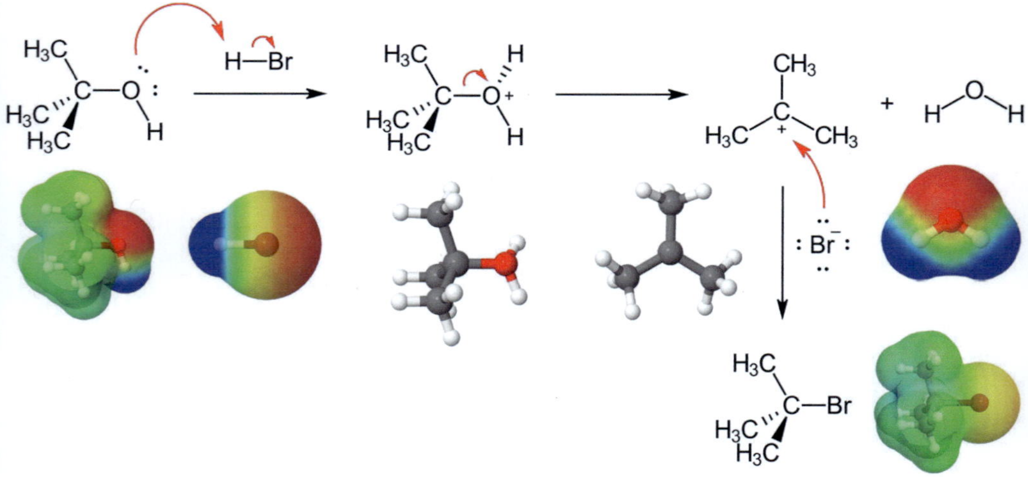

Figure 6.4 *MEP surfaces can be used in visualizing chemical reactions.*

LEARNING TASK

Study some textbooks used in your country and find the contexts in which electrostatic potential maps are used.

EXERCISE 5. PARTIAL CHARGES AND SOLUBILITY

Cis-trans isomerism of maleic and fumaric acids is an interesting visualization context. The molecules have the same chemical formula but different cis-trans structures (see figure 6.5). Because of the isomerism, the compounds have different chemical and physical properties. Maleic acid is highly water soluble and has a melting point of 408 K. Fumaric acid is poorly water soluble and has a much higher melting point of 560 K. Let's study these molecular structures with Edumol and analyze the reasons for the differences in the properties.

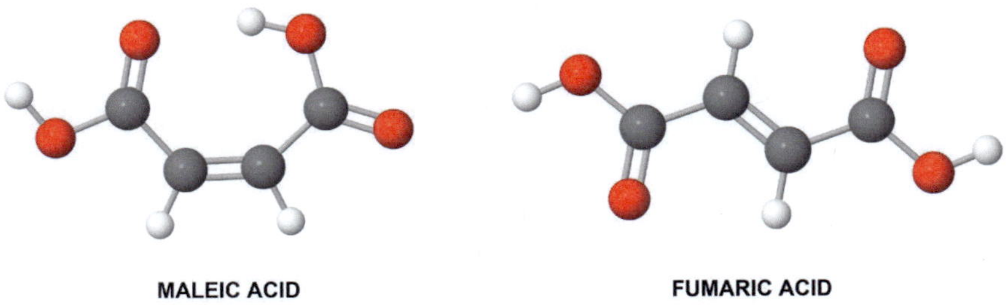

Figure 6.5 *Maleic and fumaric acid models.*

Visualize:

- Which molecule has a greater dipole moment and how is it visualized in Edumol?
  - *Files from internal database: C6 – Maleic acid and C6 – Fumaric acid*
- Why is maleic acid highly water soluble but fumaric acid is poorly water soluble?
  - *Files: C6 – Maleic acid dimer, C6 – Fumaric acid dimer, C6 – Maleic acid and C6 – Fumaric acid*
- Why does fumaric acid have a melting point of 560 K, but maleic acid a melting point of 408 K?
  - *Files: C6 – Maleic acid dimer and C6 – Fumaric acid dimer*

Answers

Maleic acid is more soluble in water than fumaric acid. Maleic acid is less stable and far more polar than fumaric acid. Fumaric acid is actually a non polar molecule because the acids are on opposite sides of the molecule; thus eliminating the dipole moment. This can be visualized with the Partial Charge tool.

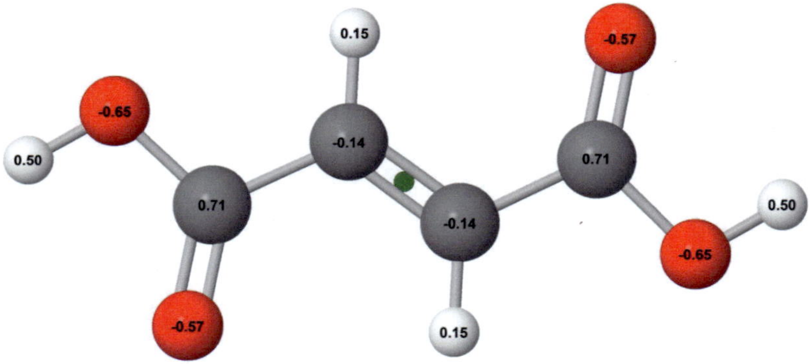

Figure 6.6 *The green dipole vector that indicates the size of the dipole moment has a length of zero in fumaric acid.*

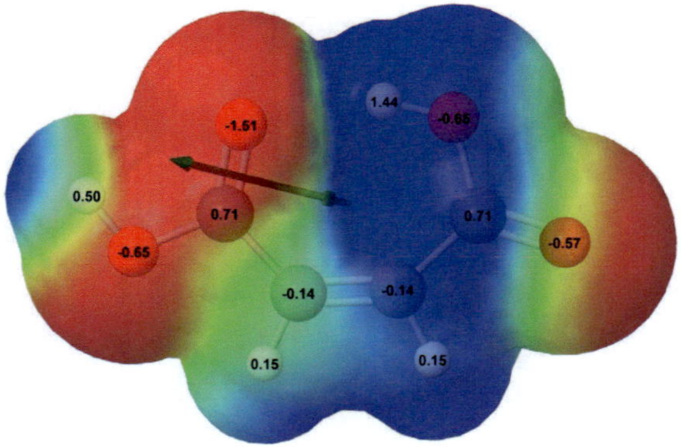

Figure 6.7 *In the maleic acid model, the dipole vector is clearly seen. The partial charges and MEP map illustrates the polarity.*

The lower melting point of maleic acid can be explained by differences in hydrogen bonding. Maleic acid molecules have intramolecular hydrogen bonding that takes place at the expense of intermolecular interactions (see figure 6.8). The shape of fumaric acid allows it to form more hydrogen bonds between other fumaric acid molecules (intermolecular bonding) (figure 6.9). Therefore, the interactions between fumaric acid molecules are greater than the interactions between maleic acid molecules. Accordingly, it takes more energy to melt fumaric acid.

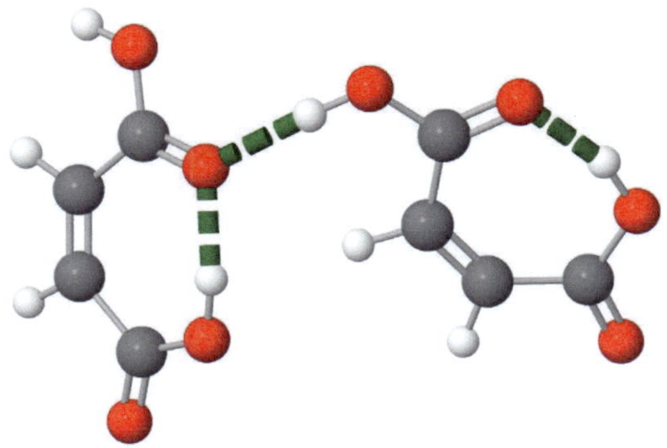

Figure 6.8 *Maleic acid dimer. Hydrogen bonds are under the Biomodel menu.*

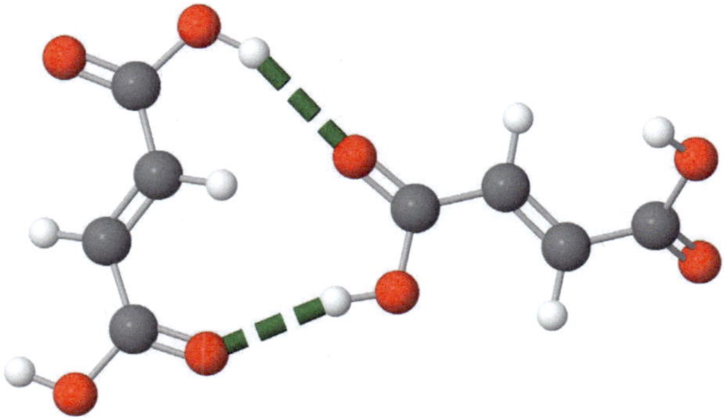

Figure 6.9 *Fumaric acid dimer has more possibilities to form intermolecular hydrogen bonds than maleic acid.*

EXERCISE 6. VISUALIZING ACID STRENGTHS

Acid strength is an important molecular property that can be visualized via electrostatic potential maps. The value of the electrostatic potential around the acidic hydrogen in a neutral acid allows the comparison of relative acidic strengths. The more positive the potential, the stronger the acid.

Electrostatic potential maps visualize trends in acidity. For example, the acidic hydrogen in nitric acid (a strong acid) is more positive than the acidic hydrogen in acetic acid (a weak acid). Likewise, the acidic hydrogen is more positive in acetic acid compared to ethanol, which is a very weak acid.

Implementation

Open three tabs in web browser and load Edumol on each tab. Search the Small Molecules database for the following molecules:

- ethanol
- acetic acid
- nitric acid.

Visualize the molecular electrostatic potential surface and rotate the models in such a way that the positive areas around the acidic hydrogen are well displayed. Strong acids clearly have more positive potential around the acidic hydrogen than weaker acids (see figure 6.11). The results can be saved as a images or files.

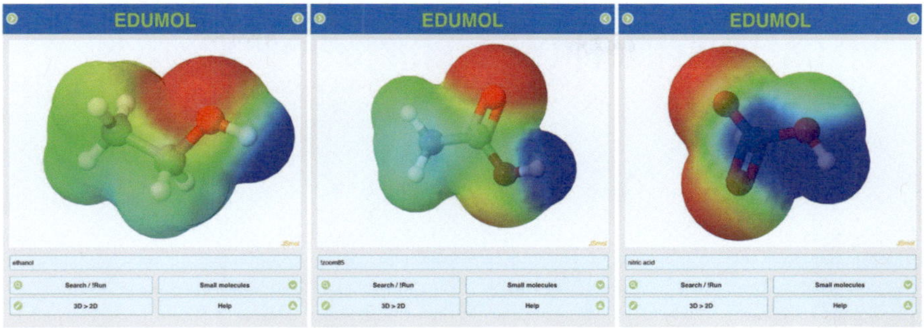

Figure 6.10 *Tip for laptop or desktop users: Open three tabs side by side in order to easily compare the models.*

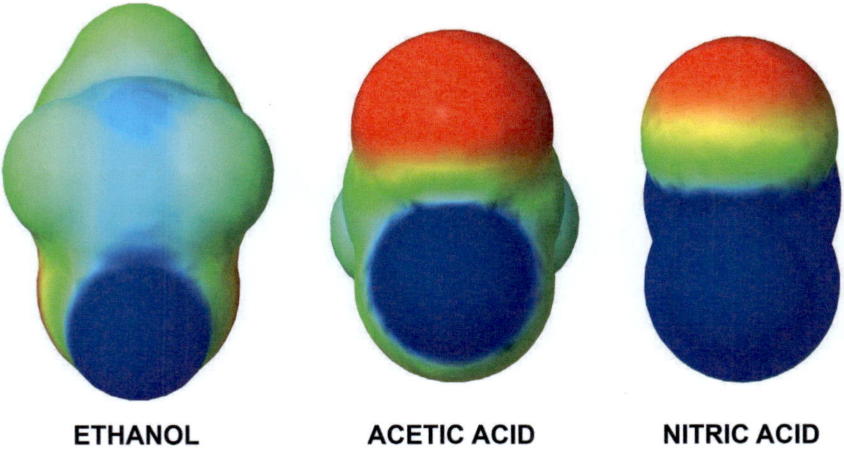

Figure 6.11 *Ethanol, acetic acid and nitric acid. Which acid is the strongest based on MEP surfaces? Images are rendered using the Export png (HQ) function.*

REFERENCE

Bottyan, T. (2015). Electrostatic Potential maps. Chemistry LibreTexts.

7. VISUALIZING CHEMICAL BONDING

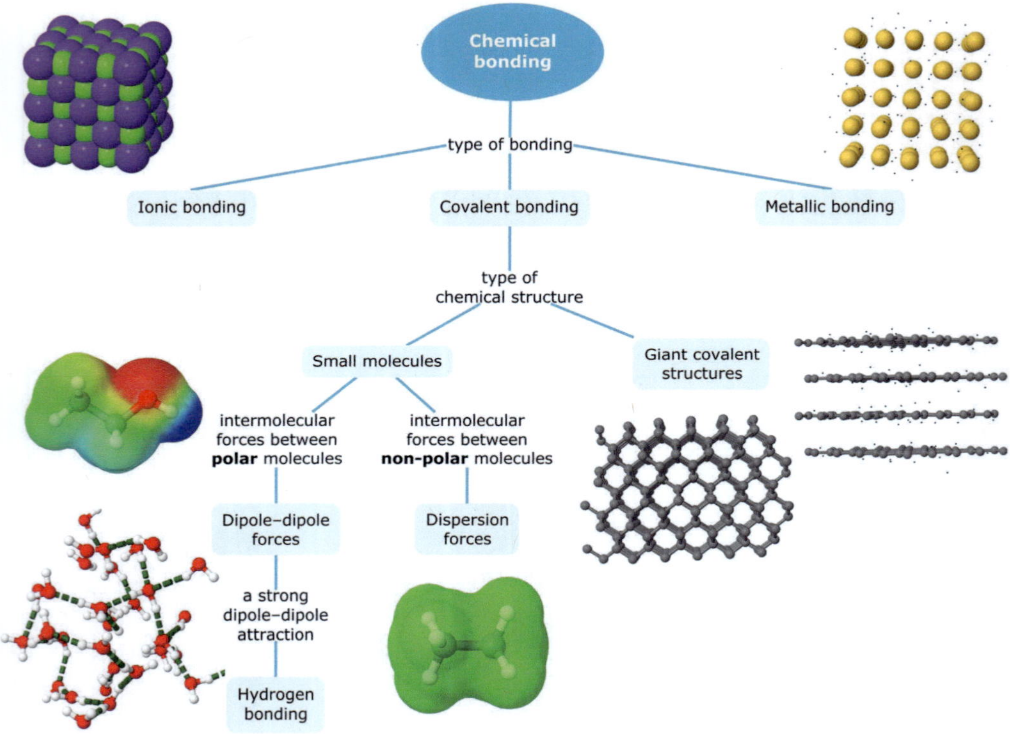

Figure 7.1 *Bonding categories and molecular forces discussed in this chapter.*

Chemical bonding is a fundamental concept in chemistry. Unfortunately though, the concept of bonding is quite complex and abstract. That is why many students find it difficult to understand chemical bonding and, as a result, develop alternative conceptions. However, the complexity of the topic is not the only reason students develop alternative conceptions.

Students' alternative conceptions are also a result of how they have been taught the concept.

Students are expected to develop an understanding of bonding based on the wide range of complex and sophisticated models used by chemists to make sense of bonding. Additionally, students are expected to interpret a variety of symbolic representations of chemical bonds. These expectations partially explain why students find the concept of bonding difficult to comprehend. In order to facilitate student understanding of bonding, the models, symbols and explanations of bonding need to be both simplistic and authentic representations of scientific concepts and models. Another major hindrance to student understanding of bonding is that students live and operate in a macroscopic world and thus find it difficult to shift between macroscopic and molecular levels.

Computer-based molecular modeling has opened various possibilities for teaching chemistry. It has helped to promote students' understanding of chemistry at four levels: the macroscopic, sub-microscopic, symbolic and chemical process levels. Computer-based visual models also help students understand chemical concepts, theories and molecular structures. Additionally, Ardac and Akaygun (2005) found that students who studied molecular representations and the structure of matter with the use of computer-based models outperformed their peers who didn't use computer-based models.

Throughout the years, vast amount of empirical research has produced ideas on how to teach chemical bonding in a way that facilitates student understanding of bonding and prevents students from developing alternative conceptions. Exercises 7, 8 and 9 utilize these research findings and computer-based molecular modeling to enhance students' understanding of chemical bonding. The exercises can be used as classroom activities or for homework, or the teacher can utilize them as demonstrations during lectures.

EXERCISE 7. INTERMOLECULAR AND INTRAMOLECULAR FORCES

Ionic bonds

Sodium chloride lattice is a classic example of an ionic bond. If sodium chloride is formed in a Born–Haber cycle, energy is required when: 1. a Cl–Cl bond is broken, 2. sodium changes state from solid to gas, 3. gaseous sodium is ionized. Energy is released when gaseous chlorine gains an electron. Yet the formation of NaCl from Na and Cl_2 is an exothermic reaction.

From the Edumol database (left panel) search for and open (on different tabs): **1. C7 – NaCl, 2. C7 – Salt crystal** and **3. C7 – Water**. Visualize the salt crystal as a CPK model as well.

Question 1. *Explain, based on electrical forces, why the formation of sodium chloride is an exothermic reaction.*

Discussion – ionic compounds have distinct properties

a) Aqueous solutions of ionic compounds conduct electricity

b) Ionic compounds are soluble in polar solvents. For example, sodium chloride is soluble in water (a polar solvent).

c) Ionic crystals are relatively hard but quite often brittle (break easily).

d) The melting and boiling points of ionic compounds are usually very high. For sodium chloride, the melting point is 801 °C and the boiling point is 1413 °C.

e) The volatility of ionic compounds tends to be very low. Volatility means the tendency of a substance to vaporize.

Question 2.

a & b) Draw your own model of what an aqueous solution of NaCl looks like based on the models of water and salt. Explain your model and also explain why aqueous solutions of ionic compounds conduct electricity.

c) Use the previously visualized CPK model of a salt crystal to explain why ionic crystals are brittle. (Hint: think what happens to the ions if you hit a crystal of salt with a knife or spoon)

d) Can you explain property d (high melting and boiling points) based on the structure of ionic compounds?

e) Can you explain property e (low volatility) based on the structure of ionic compounds?

Covalent bonds – small molecules

In a covalent bond electrons are shared between atoms. Cl_2 and HCl are both bonded covalently. Search the Edumol database for **C7 – HCl** and **C7 – Cl2**, and open them on different tabs. Visualize the MEP surfaces.

Question 3. *Explain the electron distribution of Cl_2 and HCl based on the visualization. Also explain why HCl is a polar covalent compound and Cl_2 is a non-polar covalent compound.*

Discussion – covalent bonds

Covalent bonds between atoms within a molecule are very strong. The forces of attraction between molecules are generally much weaker. These intermolecular forces depend on the polarity of molecules. Polar molecules are held together mainly by dipole–dipole intermolecular forces between molecules, and non-polar molecules are held together mainly by dispersion forces between molecules.

Dipole-dipole forces – intermolecular forces

There are always polar covalent bonds in polar molecules (molecules with permanent dipole moments). Search the Edumol database for **C7 – ICl** and visualize its MEP surface.

Question 4. *Based on the visualization, sketch the intermolecular forces between three ICl molecules and explain the electrostatic attractions.*

Discussion – hydrogen bonding

Hydrogen bonding is a strong dipole–dipole attraction. It occurs when hydrogen is bonded directly to a small highly electronegative atom such as fluorine, oxygen or nitrogen. As the electron pair is drawn away from the hydrogen atom by the more electronegative atom, all that remains is the proton in the nucleus since there are no inner electrons. The proton then attracts a pair of non-bonding electrons from another F, O or N.

Search the Edumol database for **C7 – Water**, then visualize the hydrogen bonds. Then search the Edumol database for **C7 – Ice**, and visualize the hydrogen bonds.

Question 5. *What factors affect the strength of dipole–dipole bonding?*

Dispersion forces - intermolecular forces

Even in non-polar covalent molecules (such as H_2) the electrons can be unevenly spread at any moment, producing temporary (instantaneous) dipoles. An instantaneous dipole can induce another dipole in a neighboring particle, which results in weak attraction between the two particles as seen in figure 7.2.

***Figure** 7.2 Example of an induced dipole.*

Question 6. *Explain why dispersion forces increase with increasing mass.*

***Question** 7. Simple covalent compounds have distinct properties. Explain the following based on electrostatic attractions.*

Covalent compounds have low melting and boiling points and high volatility. In their solid, liquid and gas states covalent compounds do not conduct electricity. Similarly, in solutions they do not conduct electricity.

Explain the solubility of the following covalent compounds in water, ethane and ethanol. Also discuss the solubility of heptanol. You can visualize the structure as well as the MEP surface to guide you if necessary.

*Edumol database: **C7 – Q7**.*

Giant covalent structures

Giant covalent structures contain numerous atoms joined to adjacent atoms by covalent bonds. The atoms are usually arranged in giant regular lattices. Diamond and graphite are giant covalent structures.

From the Edumol database search for and open **C7 – Diamond** and **C7 – Graphite(e)** to visualize the structure of diamond and graphite. Notice the delocalized electrons between the layers of graphite.

Question 8. *Explain based on the model why graphite is soft and slippery, has a high melting point and conducts electricity.*

Question 9. *Explain based on the model why diamond is hard, has a very high melting point (more than 3600 °C) and does not conduct electricity.*

Metallic bonding

Metals form giant structures in which valence electrons of the metal are free to move. As valence electrons in metals have become detached from the individual atoms, metals consist of positive ions and delocalized electrons.

Search the Edumol database for **C7 – Gold** to view the structure of gold as an example of metallic bonding.

Question 10. *Explain metallic bonding based on the electrostatic attractions.*

Most metals have distinct properties. The extent of their properties differs depending on the metal; for example, some are better conductors of electricity than others.

Question 11. *Explain the following properties based on the structure of metals:*

- metals are malleable and ductile
- metals have high melting and boiling points
- metals conduct electricity.

EXERCISE 8. SIMILARITIES BETWEEN IONIC, COVALENT AND POLAR COVALENT BONDING

Search the Edumol database for **C7 – NaCl, C7 – Cl_2** and **C7 – HCl**, and visualize the MEP.

Generally, it is possible to predict based on the difference in electronegativity, whether a compound has ionic or covalent bonds. If the difference in electronegativity between the elements is greater than 1.8, the compound is usually ionic. However, compounds are often somewhere between ionic and covalent.

Question 12. *Based on your knowledge of ionic bonds, polar covalent bonds, covalent bonds and the models, explain why ionic and covalent bonding should not be viewed as a dichotomy.*

EXERCISE 9. BOND LENGTH

Bond length is the average distance between the centers of the nuclei of two bonded atoms in a molecule. The shorter the bond length is, the larger the bond energy. In Edumol bond lengths are measured in picometers (pm).

Question 13. *What is the relation between the bond length and the size of the bonded atoms? Examine the bond length of F_2, Cl_2, and Br_2 to assist you in your explanation.*

Edumol database: ***C7 – Q13.***

Question 14. *What is the relation between bond length and bond order (number of chemical bonds between a pair of atoms)? Examine the bond lengths between the carbon atoms of ethane, ethene and ethyne to assist you in your explanation.*

Edumol database: ***C7 – Q14****.*

REFERENCES

Aksela, M., & Lundell, J. (2008). Computer-based molecular modelling: Finnish school teachers' experiences and views. *Chemistry Education Research and Practice, 9*(4), 301–308.

Ardac, D., & Akaygun, S. (2005). Using static and dynamic visuals to represent chemical change at molecular level. *International Journal of Science Education, 27*, 1269–1298.

Barak, M., & Dori, Y. J. (2005). Enhancing undergraduate students' chemistry understanding through project-based learning in an IT environment. *Science Education, 89*, 117–139.

Gabel, D. (1996). The complexity of chemistry: Research for teaching in the 21st century. Paper presented at the 14th International Conference on Chemical Education, Brisbane, Australia.

Harrison, A. G., & Treagust, D. F. (2000). Learning about atoms, molecules, and chemical bonds: A case study of multiple-model use in grade 11 chemistry. *Science Education, 84*, 352–381.

Johnstone, A. H. (1991). Why is science difficult to learn? Things are seldom what they seem. *Journal of Computer Assisted Learning, 7*, 75–83.

Levy Nahum, T., Mamlok-Naaman, R., Hofstein, A., & Taber, K. S. (2010). Teaching and learning the concept of chemical bonding. *Studies in Science Education, 46*(2), 179–207. Philadelphia.

Robinson, W. (2003). Chemistry problem-solving: Symbol, macro, micro, and process aspects. *Journal of Chemical Education, 80*, 978–982.

Taber, K. S., & Coll, R. K. (2002). Bonding. In J. K. Gilbert, O. De Jong, R. Justi, D. F. Treagust & J. H. Van Driel (Eds.), *Chemical Education: Towards Research-based Practice* (pp. 213–234). Springer Netherlands.

8. VISUALIZING HYBRIDIZATION

Hybridization is a theory used to explain molecular shape in organic chemistry. It is a process where atomic orbitals form new hybrid orbitals. Hybrid orbitals settle into a configuration where the molecule has the lowest energy.

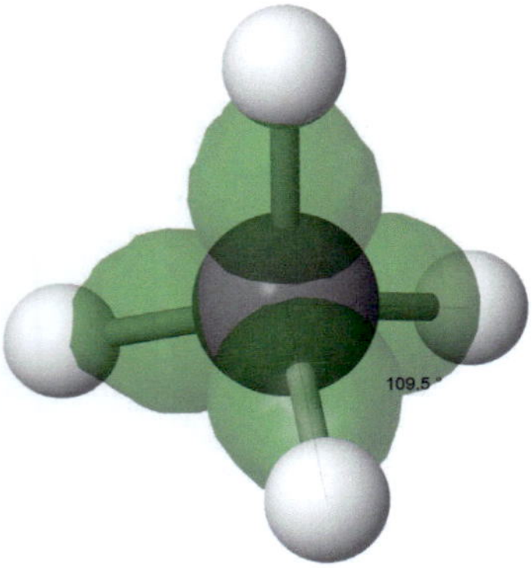

Figure 8.1 *Tetrahedral shape.*

For example, sp^3 hybridization explains the tetrahedral shape of molecules (see figure 8.1). To form four sp^3 orbitals, one 2s orbital and three 2p orbitals need to hybridize. Thus each sp^3 hybrid orbital has 75% p character and 25% s character. The sp^3 orbitals are oriented at a bond angle of 109.5°, which leads to tetrahedral geometry and lowest electronic energy.

In chemistry education, hybridization is considered one of the most challenging topics. According to research literature, students may produce many alternative concepts related to the topic. Alternative conceptions of hybridization have been studied by e.g. Nakiboğlu (2003).

In her paper, Nakiboğlu suggests ideas to consider when teaching hybridization. The paper is freely available on the internet. We highly recommend downloading and reading it.

In the exercises related to hybridization we have taken into account one of her ideas, namely to explore the hybridization of not only carbon but other elements as well.

EXERCISE 10. HYBRIDIZATION OF CARBON ATOMS

In this exercise we visualize the classic textbook hybrid orbitals using Edumol command line tools. Hybrid orbitals are not incorporated into Edumol user interface by default, which means they must be visualized using JSmol scripting commands. Some of the commands are a bit long, so it is recommended to do all the hybridization exercises using a laptop, a desktop computer or a tablet with a manual keyboard.

Hybrid orbitals are rendered via LCAOCARTOON command. It displays them as cartoonish opaque or translucent shapes.

The documentation can be read on:
http://chemapps.stolaf.edu/jmol/docs/#lcaocartoon.

Warming up

First select the atoms, set lcaocartoon on, set color, choose display mode for the desired hybrid orbital and write the orbital name inside quotation marks.

```
methane
!select carbon; lcaocartoon color red translucent create "s"
!select hydrogen; lcaocartoon color red translucent create "s"
```

And this is how orbitals are taken off: `!select all; lcaocartoon off`

sp^3

Let's continue to work with methane and visualize its sp^3 hybridization.

```
methane
!select all; lcaocartoon color green translucent create "sp3"
```

The COLOR option sets the color for the orbital. The default color is gamboge yellow. TRANSLUCENT makes it transparent. The default value is solid.

It is also possible to visualize one orbital at a time. For example for the carbon atom:

```
!select carbon; lcaocartoon "sp3a"
!select carbon; lcaocartoon "sp3b"
!select carbon; lcaocartoon "sp3c"
!select carbon; lcaocartoon "sp3d"
```

TIP: *Bond angles are essential. If bond angles need to be measured, the ball-and-stick model is the best option because the atoms are big enough to be selected easily.*

Finally, add the s orbitals of hydrogen to the visualization:

```
!select hydrogen; lcaocartoon color red translucent create "s"
!select all; lcaocartoon color green translucent create "sp3"
```

If the translucent value is left out, hybrid orbitals are rendered as solid shapes.

```
ethane
!select all; lcaocartoon color green create "sp3"
```

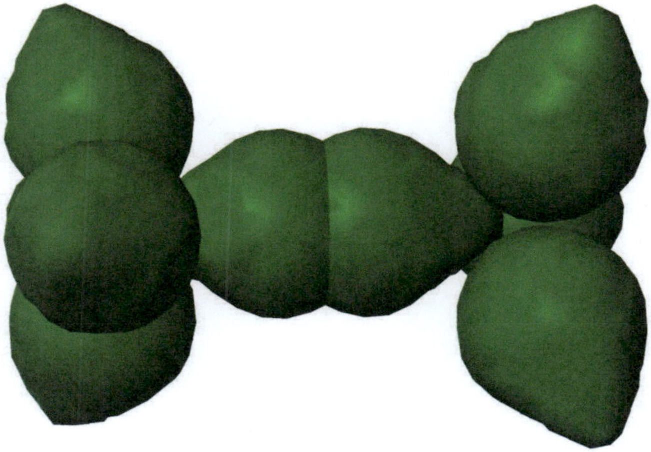

sp^2

First create a pi bond in the same plane so that the double bond is visualized. Axes help to determine the direction of the plane. Then add three sp^2 orbitals.

```
ethene
!lcaocartoon color green translucent create "pz"
!lcaocartoon color blue translucent create "sp2a" "sp2b" "sp2c"
```

Add the s orbitals.

```
!select hydrogen; lcaocartoon color red translucent create "s"
```

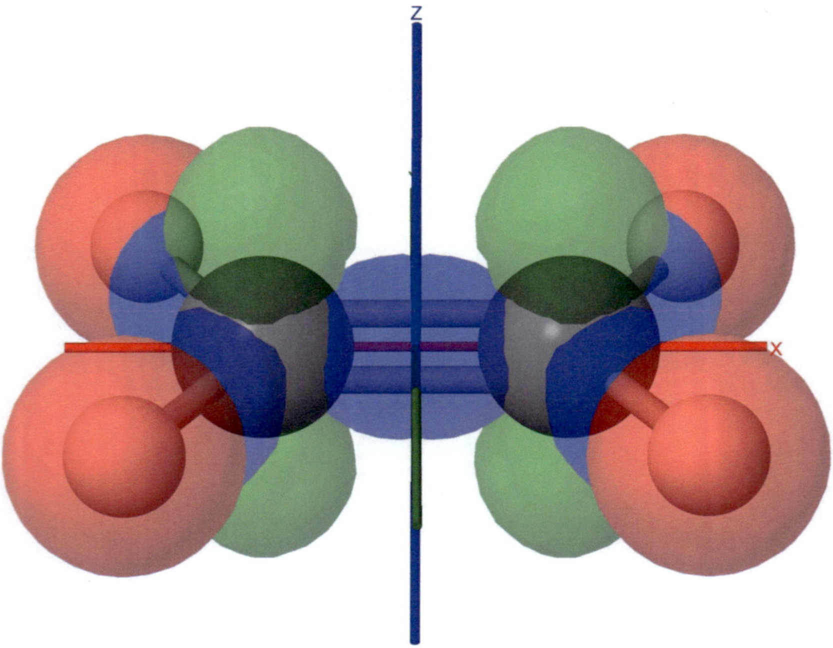

sp

```
ethyne
!lcaocartoon color green create "px" color red "py" color
purple "spa" "spb"
```

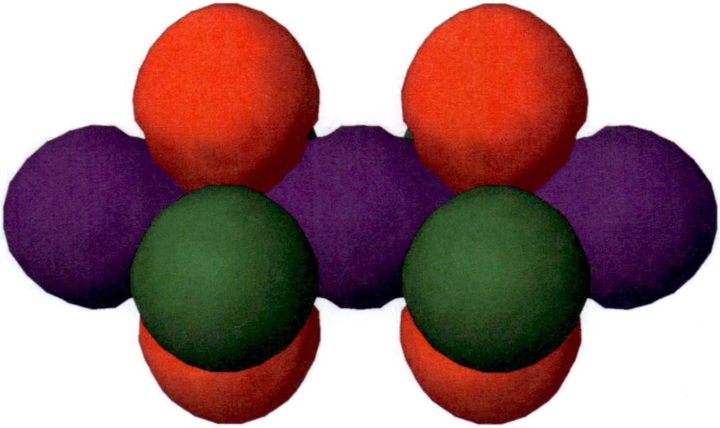

EXERCISE 11. HYBRIDIZATION OF NITROGEN AND OXYGEN

Nitrogen

In ammonia molecules, nitrogen has one lone pair. It can be visualized as a lobe using CREATE command or as two dots using LONEPAIR.

```
ammonia
!select nitrogen; lcaocartoon create "sp3a" "sp3b" "sp3c"
!lcaocartoon lonepair "lp"
```

Oxygen

In water molecules, oxygen atom has two lone pairs which can be visualized using LBA and LPB values.

```
water
!select oxygen; lcaocartoon create "sp3a" "sp3b"
!lcaocartoon color blue lonepair "lpa" "lpb"
```

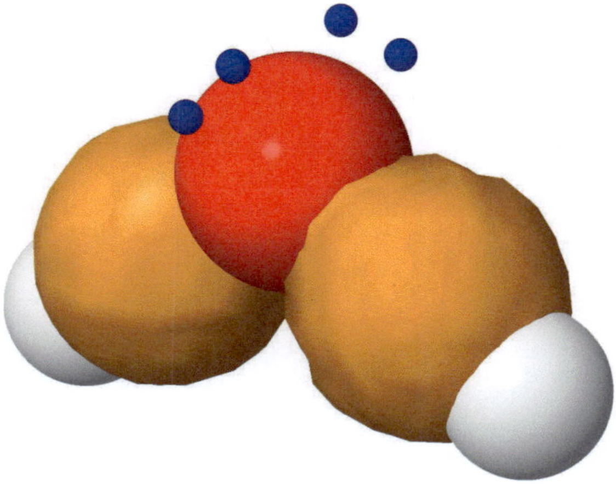

EXERCISE 12. HYBRIDIZATION OF BENZENE

Use commands learned in the previous exercises and visualize the hybridization of benzene.

Example model in Edumol database: **C8 – Benzene**.

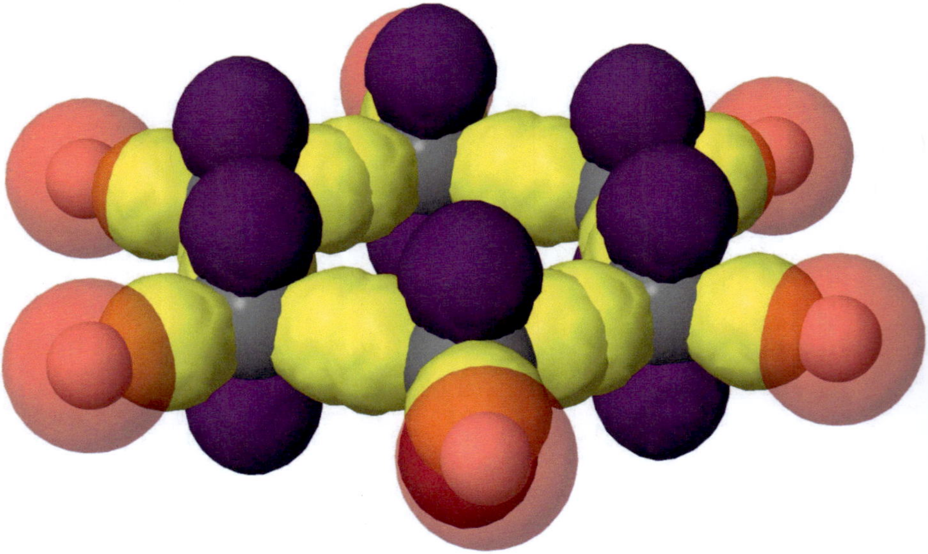

REFERENCES

Chemwiki. (2014). Hybrid Orbitals. Retrieved 26.7.2016 from http://chem.libretexts.org/Core/Organic_Chemistry/Fundamentals/Hybrid_Orbitals.

Nakiboğlu, C. (2003). Instructional misconceptions of Turkish prospective chemistry teachers about atomic orbitals and hybridization. *Chemistry Education: Research and Practice* 4(2), 171–188. <http://www.chem.uoi.gr/cerp/2003_May/pdf/06Nakiboglu.pdf>

9. ISOMERISM

Isomers are molecules with the same molecular formula but different structural or spatial arrangements of atoms, resulting in different chemical and physical properties. Isomers can be sorted in several categories (see figure 9.1); these can be visualized with Edumol.

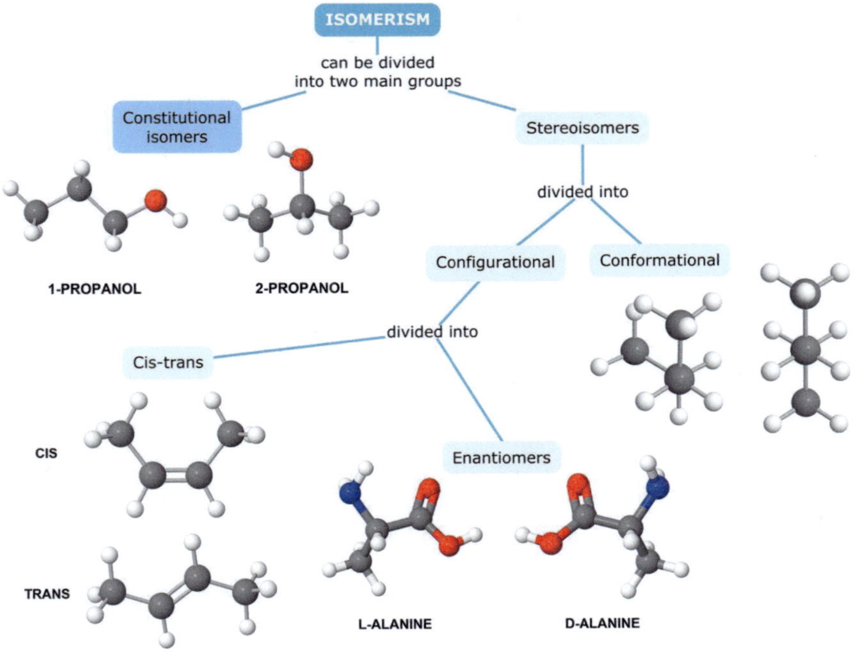

***Figure 9.1** Isomerism categories studied in the chapter.*

Edumol can only perform molecular mechanics calculations. Molecular mechanics uses force fields to calculate the steric energy of the molecule and then adjusts the molecule into a local minimum. This means that Edumol can be used to compare energies of isomers which differ only by the spatial arrangement of atoms and not by the order of atoms. So comparisons are meaningful for conformational and cis-trans isomers but not for constitutional isomers. However, constitutional isomers can always be visualized.

EXERCISE 13. VISUALIZING CONSTITUTIONAL ISOMERISM

Choose an alcohol such as butanol or pentanol. Sketch one isomer on the 2D side and switch to the 3D side. Check the molecular formula in the NIST search box on the left panel. Then switch back to the 2D side and modify the molecule to make it another isomer and switch back again. Confirm that the molecular formula has remained the same.

EXERCISE 14. GEOMETRICAL ISOMERISM AND ENERGY

Retrieve cis-2-butene from the database, minimize it, calculate the energy and record it. Repeat the same for the trans-2-butene (see figure 9.2). Which isomer is lower in energy?

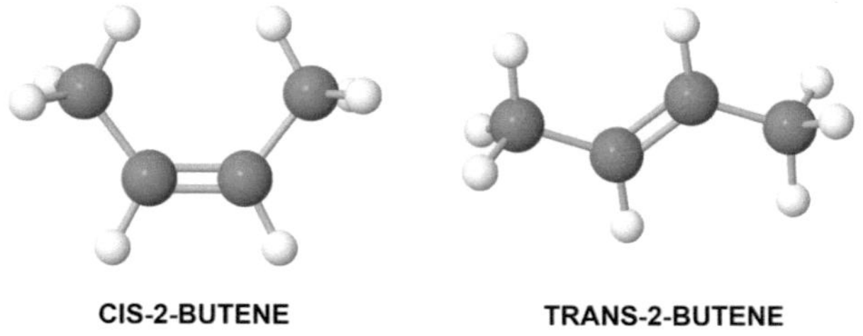

Figure 9.2 *Notice that both structures have more than one local minimum structure. These are the global minimums with minimal steric hindrance.*

EXERCISE 15. ENANTIOMERS

Here are a couple of example molecules tovisualize enantiomers. This exercise is much easier if carried out with a device that supports split screen mode. It is handy to have the models visible simultaneously for comparisons.

Alanine

Alanine is a simple optically active amino acid. Search for the structures from Small Molecules database or from PubChem. Use the search terms L-alanine and D-alanine (see figure 9.3).

Notice that the amino group in L-alanine needs to be adjusted slightly in order to get the exact mirror images of the molecules. Use Drag and Minimize tool in the Calculate menu on the right panel of Edumol user interface.

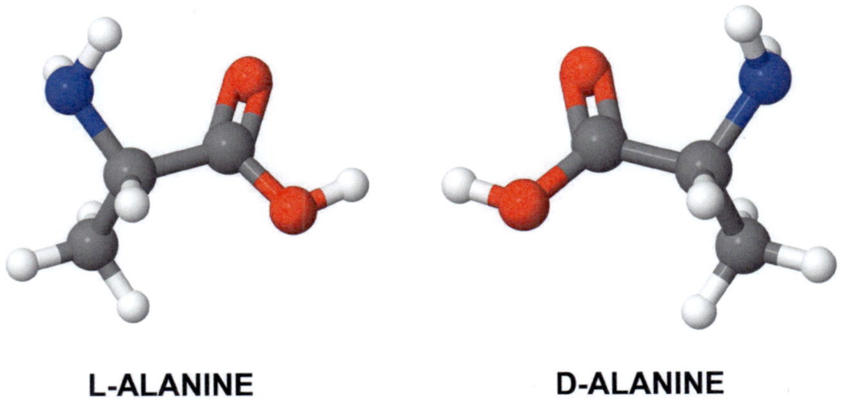

Figure 9.3 *L-alanine and D-alanine models.*

Thalidomide

Thalidomide is a classical molecule from the history of medicinal chemistry. It was designed to help people sleep and reduce nausea during pregnancy but, as was later revealed, it caused birth defects, and infants were born with shortened limbs. The drug consisted of two thalidomide enantiomers. The *S*-enantiomer was teratogenic whereas the *R*-enantiomer was an effective sedative.

Visualize the two thalidomide isomers. Search PubChem database for (*R*)-thalidomide and (*S*)-thalidomide (see figure 9.4), rotate the molecules and change the model types. Can you find the difference?

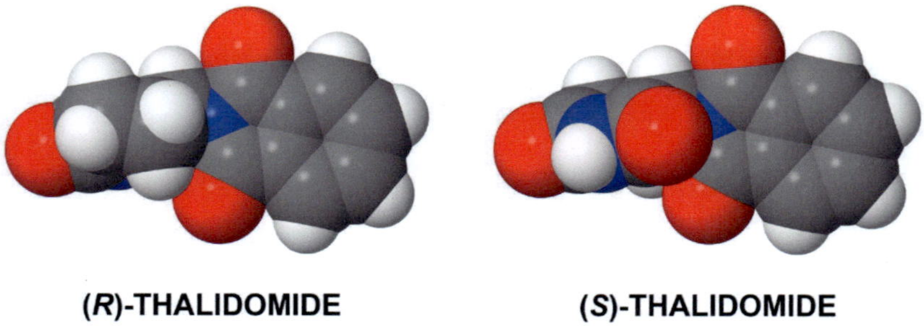

Figure 9.4 *(R)-thalidomide and (S)-thalidomide models.*

Extra – Other interesting molecules

Think of some other optical isomers that would interest your students. Prepare a visualization task sheet for them using Edumol application.

EXERCISE 16. CONFORMATIONAL ISOMERISM

This exercise is a bit more challenging. Search for or sketch a butane molecule, visualize all conformers, take a picture of each one and calculate the enthalpy differences (see figure 9.5).

Visualizations can be carried out via the following process:

- Rotate the molecule model into a good position.
- Measure the torsion angle of the four carbon atoms.
- Tap the Drag and Minimize tool to activate it.
- Finally tap the Calculate energy feature to switch it on, and Edumol will calculate the energy of the molecule.
- Then try to adjust the torsion angle to about 0, 60, 120 and 180 degrees, take a picture and record the energy values. Be patient, this can be very tricky.

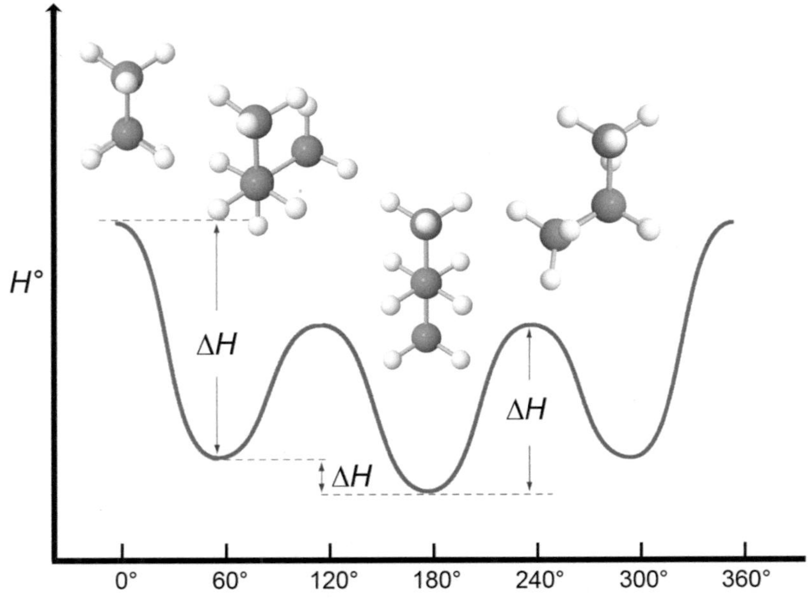

***Figure 9.5** Conformations of butane.*

10. INTRODUCTION TO CHEMINFORMATICS: BIOMODELS AND DATABASES

Cheminformatics is a broad research field covering all aspects of the representation and use of chemical and related biological information on computers. This chapter offers a short introduction to cheminformatics and illustrates how Edumol takes advantage of the fundamental rules of the field.

THE HISTORY OF CHEMINFORMATICS

Chemists were pioneers in computer-aided research. Computers have been used in chemistry research since the 1950s when chemists began to develop statistical chemistry models. The next big step was computer-based molecular graphics which emerged in the 1960s. In the 1970s and 1980s the use of computers expanded rapidly into the field of computational drug discovery research.

The growth of cheminformatics has not stopped. On the contrary, at this age of digitalization, cheminformatics is a vital multidisciplinary field of chemistry. Traditional areas of applications are databases (e.g. chemicals and patents), pharmaceutical research and drug discovery, journal article indexing and integrations to other fields like bioinformatics.

Read more about cheminfomatics from learncheminformatics.com.

Success stories of cheminformatics and Edumol

Cheminformatics has achieved a lot in traditional research areas. The biggest success stories of cheminformatics are

1. searching databases for chemical structures

2. representing and visualizing 2D and 3D structures with computers.

Both of these features are very well covered in Edumol. The core of Edumol is JSmol, which offers integrations to multiple databases, NCl (small molecules), Protein Data Bank (macromolecules) and PubChem. This means that structures can be retrieved into Edumol in multiple ways. For example, small molecules can be retrieved from NCI database via

- common name: acetylsalicylic acid or aspirine
- systematic name: 2-(acetoxy)benzoic acid
- CAS Registry Number: 50-78-2 (see figure 10.1)
- SMILES: CC(=O)OC1=CC=CC=C1C(=O)O
- InChI: InChI=1S/C9H8O4/c1-6(10)13-8-5-3-2-4-7(8)9(11)12/h2-5H,1H3,(H,11,12).

Figure 10.1 *Search by CAS Registry number.*

VISUALIZING BIOMODELS

After retrieving the data into Edumol system, it must be visualized. In earlier chapters we have discussed Edumol's 2D modeling and 3D visualization tools and covered topics such as building and visualizing small molecules in different model types (ball-and-stick, CPK, stick and wire). In this chapter we discuss increasing the size of the molecules and visualizing biomodels.

Retrieving biomodels

Edumol is integrated into the Protein Data Bank, where the structures must be retrieved using PDB's own PDB ID entry code. Therefore, you have to search the PDB database to get the code in order to find the same model in Edumol. For example, the ID for DNA is 1BNA and for a protein called melittin it is 2MLT.

Just select Macromolecules database and try those codes out.

Biomodel menu

After retrieving the data it can be visualized in Edumol via tools from the Biomodel menu.

- **Backbone** renders the protein backbone or nucleic acid helix and displays disulfide bonds, if present, along the backbone as thick yellow tubes.
- **Ribbons** render the protein backbone or nucleic acid helix using a flat band.

- **H-bonds** toggles hydrogen bonds on and off. Hydrogen bonds are visualized as dashed green lines. Thickness of the H-bonds can be changed via command !hbonds x.x. Default value 0.15.
- **SS-bonds** visualize disulfide bridges as yellow lines. SS-bonds are present in e.g. 5E5E. The thickness of SS-bonds can be changed via command !ssbonds x.x. Default value is 0.4.
- **Color amino** renders 20 essential amino acids in certain colors. Backbone or Ribbons must be switched on. Color codes can be found on the Jmol site. (http://jmol.sourceforge.net/jscolors)
- **Structure** renders different types of protein secondary structures: three types of helices, beta strands or sheets, turns and loops (see table 10.1).
- **Purple** renders protein structures in a purple color.
- **Water** selects and shows water residues as a CPK model.
- **Ligands** select and show ligands as a CPK model.

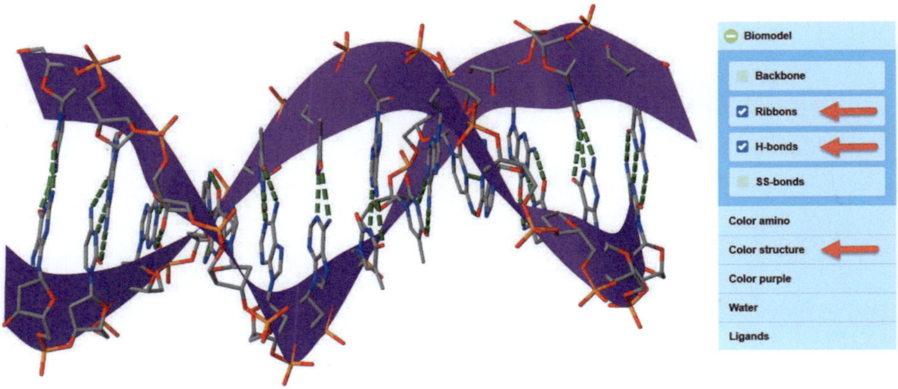

Figure 10.2 *DNA (1BNA) in a stick model, protein ribbons, H-bonds and color structure.*

EXERCISE 17. RETRIEVING DATA IN DIFFERENT WAYS

Try out different methods to get molecular data into Edumol. Select Small Molecules database (NCI), choose a compound and test retrieving data via

- common name
- systematic name
- CAS Registry Number
- SMILES
- InChI.

While testing different retrieving methods, think about the ways in which you could use this feature in your teaching.

- Could this be a way to teach systematic names?
- Can the CAS Registry Number retrieving be applied in research projects or laboratory exercises?

EXERCISE 18. VISUALIZING BIOMODELS

Choose a protein, find out its PDB ID from the Protein Data Bank and retrieve it into Edumol via Macromolecules database. Try to visualize different protein structures and save them as images or files for later use.

Primary structure: Select the best model type for primary structures and zoom in. This is the individual amino acid level. Rotate the model and find peptide bonds, hydrogen bonds and disulfide bridges.

Secondary structures are best to illustrate using backbone and ribbon selections. Wireframe, tube and CPK models are also interesting. Try different coloring options.

Atoms in α-helices and β-sheets can also be selected and colored separately. This can come handy in some educational situations.

```
!select helix; color green; color bonds green
!select sheet; color blue; color bonds blue
```

Atoms in RNA and DNA, too, can be selected and colored:

```
!select nucleic; color red
```

Notice! Large macromolecules can't be edited on the 2D side. These models have too many atoms; thus editing sometimes crashes the application. 2D editing canvas prints a following warning:

"ERROR – problems in reading/processing MOL input."

Compare possibilities and limitations

While you are trying out different visualizations, try to think of their possibilities and limitations from an educational perspective. Add your thoughts into the table 10.1.

TABLE 10.1 POSSIBILITIES AND LIMITATIONS OF BIOMODELS

Visualization	Possibility	Limitation
Wireframe and tube	Shows atom level structures	Difficult to see secondary structures
CPK	3D volume and shape	It is not possible to see secondary structures
Backbone	Secondary structures are clear	Primary structures not visualized
Ribbons	Secondary structures are well displayed	No structural details

EXERCISE 19. MOLECULE OF THE MONTH ARTICLES

Protein Data Bank offers extremely good macromolecule visualizations in the Molecule of the Month series (http://pdb101.rcsb.org). Get familiar with the article series and choose a molecule. Try the in-service visualization tools and try to reproduce similar visualization with Edumol application.

- Retrieve the model via PDB ID.
- Use menus and additional commands in preparing the visualization.
- Save your work as an image or a file for later use.

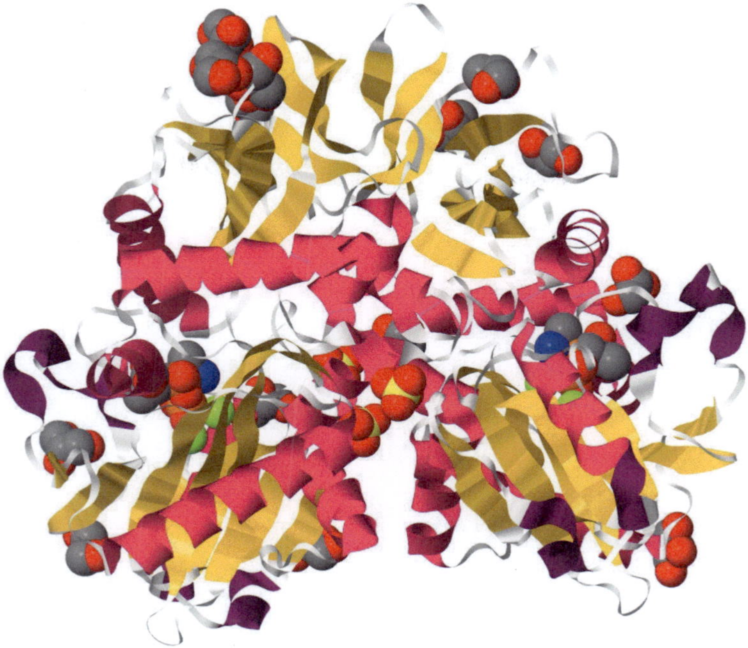

Figure 10.3 *Vancomysin, PDB ID 1E4E.*
Edumol options: Model > Hide, Biomodel > Ribbons & Ligands.

LEARNING TASKS

1. How can this article series be used in education?

2. Why would it be useful to be able to reproduce these visualizations in Edumol? For example, what do you think about the resolution of images that users can export from the applet embedded in articles?

REFERENCE

Wild, D. (2013). *Introducing Cheminformatics: An intensive self-study guide.* Self-published.

11. NEXT STEP: DESIGN YOUR OWN VISUALIZATION EXERCISES

If your chemistry teaching has educational needs that Edumol application does not cover, you can always build a specific visualization environment that addresses them. JSmol is free for everyone and easy to use. The full description of the application can be found on the **Jmol wiki site**.

http://wiki.jmol.org/index.php

We also suggest you visit other JSmol websites. The list of Jmol and JSmol websites can be found on the Jmol wiki pages. Note that some of the websites are still using Java-based Jmol which does not work on mobile devices.

http://wiki.jmol.org/index.php/Websites_Using_Jmol

EXERCISE 20. OTHER JSMOL SITES

Study the list on the Jmol wiki pages. Visit some sites and bookmark those suitable for your teaching. Pay attention to the possibilities and limitations.

Here are some excellent sites:

- Molview is one of the best sites for higher education. (molview.org)
- MolCalc – The Molecule Calculator: A Web Application for

Fast Quantum Mechanics-Based Estimation of Molecular Properties. (molcalc.org)

- Chemagic.org has an extremely versatile JSmol-based modeling kit. (chemagic.org)
- ChemEd DL 360 models is a chemical education digital library which allows users to embed models on their own website. (www.chemeddl.org)

EXERCISE 21. JSMOL WEB DEVELOPMENT

Let's finish this set of exercises by building a small JSmol site. Download JSmol from the Sourceforge: http://sourceforge.net/projects/jmol.

JSmol files are inside the Jmol package. Just unzip Jmol and then unzip JSmol.

Your first JSmol page

Begin your JSmol developer career by copying the source code of an example html file: view-source:http://edumol.fi/jsmol/ui-test.html

We have explained the vital parts of the code in comments so that you can understand the basics of JSmol coding. For more details, read the JSmol file explanations from the Jmol wiki page.

Open the page in Mozilla Firefox, Chrome or Safari. Save the file or copy & paste the source code into a text editor. For example, Notepad++ is a decent text editor that can be recommended.

Next, save the file in the unzipped JSmol folder and open it in a web browser. Now you have set a local application development environment and you should see the the following user interface (see figure 11.1).

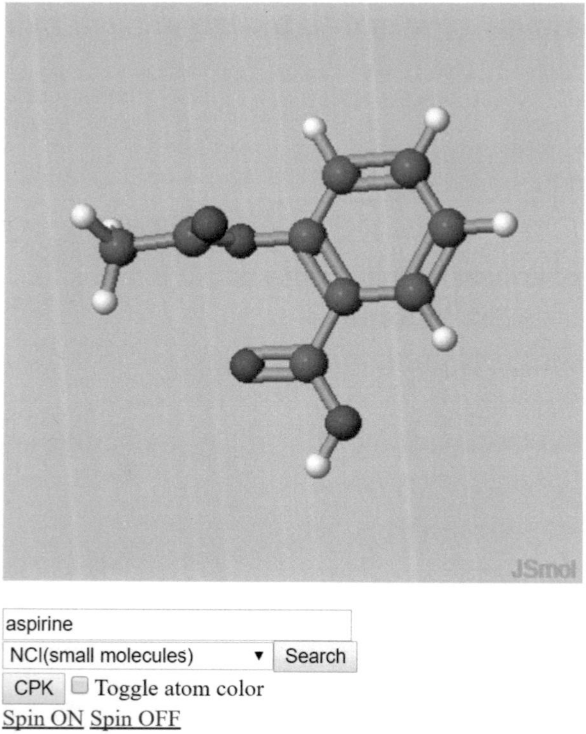

***Figure 11.1** User interface for the first JSmol page.*

Build your own user interface

JSmol is a powerful tool for building educational environments because of its user interface tools. The learning activity can be carefully constructed using checkboxes, links, buttons and radiogroups.

Here are some examples for you to try out; the full description of the tools is listed on the JSmol functions wiki page.

Checkboxes

Checkbox is a tool with two functions, such as on and off. In the example form 2 below, a model is set to spin. Always use form tag with checkboxes. It clears the checkbox selections when the page is refreshed.

```
SYNTAX
Jmol.jmolCheckbox(applet-name,script1, script0, text,
ischecked)
```

```
EXAMPLE FORM 2
<form><!-- Comment: FORM tag sets checkbox off when page reloads
-->
<script type="text/javascript">
Jmol.jmolCheckbox(jmolApplet0,"spin on","spin
off","Spin", false);
</script>
</form>
```

Let's test this. Add the example form 2 element above into the source code in the body tag below the example form 1 (line 47), save the file and refresh the page in the browser. If you see a new checkbox you have successfully modified the code. Well done!

Next just start building your own teaching settings. All the commands that you can apply in the user interface are documented in the Jmol/JSmol Interactive Scripting Documentation (https://chemapps.stolaf.edu/jmol/docs). We suggest you save it in your browser bookmarks. You are going to need it.

Buttons

Button function renders a button into the page. When the button is pressed it executes the task that it is programmed to do.

```
SYNTAX
Jmol.jmolButton(my applet, script, text)
Buttons have to be placed inside <script type="text/
javascript"> </script>
```

```
EXAMPLE
<script type="text/javascript">
Jmol.jmolButton(jmolApplet0,"CPK 30%; wireframe 25",
"Ball-and-stick");
jmolBr() // New line
Jmol.jmolButton(jmolApplet0,"write ?.png", "Save
image");
</script>
```

Links

Links work with the same logic as buttons, but it is not necessary to place them inside the <script> tag.

```
<a href="javascript:Jmol.script(jmolApplet0, 'zoom
200')">Zoom in</a>
```

EXERCISE 22. DESIGNING MOLECULAR VISUALIZATION EXERCISES

Now that you have studied the basics of molecular modeling in chemistry instruction, the next step is to design more detailed chemistry visualization exercises. Take a pen and paper or your notebook and start by writing down your specific visualization needs:

- chemistry topic
- learning goal
- needed JSmol functions
- texts and images.

Here you can find everything you can do with JSmol: https://chemapps.stolaf.edu/jmol/docs.

After planning, just start coding. The project can be iterated from the ui-test.html page. When the first version is ready, show it to your peers and test it with students. Take notes and ask for feedback.

LEARNING TASKS

1. How do you feel about developing your own sites? Do you need these skills, or does Edumol or other already existing sites fulfill all your chemistry educational needs?

2. Can the development of such sites be exploited in education in general? For example, a project work that combines art (user interface design and CSS), chemistry and computer sciences (HTML5, JavaScript and CSS).

AUTHORS

JOHANNES PERNAA, PH.D., B.E.

holds a Ph.D. in chemistry education (2011, University of Helsinki). Johannes is an entrepreneuer who is interested in all kind of ICT in chemistry education, especially molecular modeling and visualization. Johannes is also the main developer of the Edumol project. Johannes wrote **chapters 2–6 and 8–11**.

MAIJA AKSELA, PH.D.

is a Professor and the Research Director of the Unit of Chemistry Teacher Education, Department of Chemistry, University of Helsinki. She is also the Director of the LUMA (STEM) Centre of Finland and the LUMA Centre in the University of Helsinki. Maija has over 30 years' experience of chemistry education. She has published over 330 papers in the field. Maija wrote the **chapter 1**, which is the theoretical backbone of the book.

SHENELLE PEARL GHULAM, MSC.

is a chemisty teacher from Finland. Shenelle wrote her master's thesis on molecular modeling, and she wrote **chapter** 7 based on the thesis. Shenelle's thesis was supervised by Maija and Johannes.

INDEX